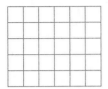

KT-441-291

CONTENTS

ELECTRICIAN'S
CALCULATIONS MANUAL

LEARNING
SUPPORT
SERVICES

Please return
on or before
the last date
stamped below

City College
NORWICH

The McGraw·Hill Companies

Library of Congress Cataloging-in-Publication Data

Fowler, Nick.
    Electrician's calculations manual / Nick Fowler.
        p.   cm.
    Includes index.
    ISBN 0-07-143654-5
    1. Electric engineering—Problems, exercises, etc.   2. Electric
engineering—Mathematics—Handbooks, manuals, etc.   3.
Mathematics—Formulae—Handbooks, manuals, etc.   I. Title.
TK151.F74   2004
621.319'2—dc22   2004057930

Portions of this work are reprinted with permission from NFPA 70-2005, *National Electrical Code®*, Copyright © 2004, National Fire Protection Association, Quincy, MA 02269. This reprinted material is not the complete and official position of the NFPA on the referenced subject which is represented only by the standard in its entirety.

1 2 3 4 5 7 8 9 0   DOC/DOC   0 10 9 8 7 6 5 4

ISBN 0-07-143654-5

The sponsoring editor for this book was Stephen S. Chapman and the production supervisor was Sherri Souffrance. It was set in Minon by Ampersand Graphics, Ltd. The art director for the cover was Anthony Landi.

Printed and bound by RR Donnelley.

McGraw-Hill books are available at special quantity discounts to use as premiums and sales promotions, or for use in corporate training programs. For more information, please write to the Director of Special Sales, McGraw-Hill Professional Publishing, Two Penn Plaza, New York, NY 10121-2298. Or contact your local bookstore.

This book was printed on recycled, acid-free paper
containing a minimum of 50% recycled, de-inked fiber.

PREFACE

The electrical industrial industry is constantly changing. It used to be that electricians could get by on their experience or what they knew before, but not today. As the electrical industry changes, so must we. In order to stay at the top of our trade, we must build on experience, knowledge, and theory. If we relied solely on experience, we would only be able to install what we were familiar with. With experience, knowledge, and theory, we can build on what we know with new technology and, more importantly, be able to understand what we are installing and why. Instead of merely hooking up a contactor and a motor, we may be called upon to install and program a variable speed drive for the motor. Fire alarm systems were fairly simple once—10 smoke detectors per zone, pull stations, pressure switches, and tamper switches and that was it. Now, we have smart detectors that communicate with the panel and can tell the panel what's going on. With inputs and outputs (I/O ports), valves can be closed or opened, doors shut, elevators recalled and locked, and alarms, both visual and audio, activated, all automatically through the fire alarm panel. Security is another growing concern that requires knowledge and experience. LANs (local area networks) are tying several electrical components to a computer to monitor a wide variety of electrical equipment. Without the fundamentals of how these systems work, we electricians will be hard pressed to install these devices, let alone trouble shoot them.

This book is intended to help the practicing electrician review the basic calculations needed for the job, to supplement the continuing education required in most states, and to aid in preparing for the tests that we sometimes need to take to take before taking work in another state. As our electrical trade changes and gets more sophisticated, so must we. Opportunities can be missed because an electrician doesn't understand sophisticated product instructions and/or wiring diagrams. More importantly, damage to equipment could result from wrong wiring.

Our trade is already too broad for an individual to know everything, even after a lifetime of working with electricity. It should give the reader some consolation that the basics of electricity stay the same, even though more sophisticated applications are being used every day.

It is also my firm belief that electricians should help each other. I was trained by older electricians and have benefited from their experience. Over the course of the years, I have met many electricians who have helped me as I was helping them. It is important that the older electricians help pass on this honorable trade to a new generation of eager and talented young people, so that they can carry on this business of electrical installation and troubleshooting. If they build on the formal training they receive and benefit from the experience of older electricians, as well as keep up with the fundamentals, we will be leaving the electrical trade in good and capable hands. If this book helps in accomplishing that goal, then I will have done my job.

Nick Fowler

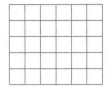

ELECTRICITY—AN OVERVIEW

Electrons are forced over a conductor at a controlled rate. These electrons are generated in cycles. Commonly in the United States, electricity is generated at 60 cycles per minute. It travels on a sine wave up a half cycle and down a half a cycle. The root mean square (rms) of the sine wave is what a voltmeter measures. For example, a 120 volt circuit travels up 120 volts and down 120 volts. This is the average voltage or rms voltage. This cycle is repeated 60 times a minute. The voltage, although it averages 120 volts, actually goes higher and lower in the peak-to-peak voltage. The peak-to-peak voltage is 1.414 times the rms voltage, yielding 169 volts peak to peak.

To complicate matters, transient, mini voltage spikes that last a millisecond occur. This can happen as a motor cycles on, or if a power surge occurs. These fluctuations can cause the voltage to spike as high as 300 volts on a 120 volt circuit. These spikes are of short duration, usually lasting a millisecond, and usually cause no problems. However, these fluctuations can and do cause problems with sensitive electronic equipment such as computer chips in a variety of applicationss, especially when they are transferring data at high speed. There are a number of solutions to limiting these spikes, including the use of voltage suppressors, isolation grounds, and lighting arrestors.

GENERATION, TRANSMISSION, AND DISTRIBUTION

From generation plants to homes and everything in between, electricity affects everyone directly and indirectly. Millions of people are employed

1

in a variety of jobs creating, distributing, and installing electrical components. Millions more are involved in manufacturing and selling everything from wire, insulators, transformers, switches, outlets, steel towers, fasteners, connectors, and light bulbs, to mention a few electrical components.

A brief explanation of what electricity is and how it is created would be helpful at this point. Remember, this explanation is simple and more detailed explanations can be found elsewhere.

Most electricity (up to 95%) in this country is generated at 13,800 volts. It can be generated in a variety of ways. Hydroelectric power uses water to turn turbines that produce electricity. Steam is another method used to turn turbines. Natural gas, nuclear power, and coal are also used to create electricity. Wind turbines and solar panels have limited use. In the future, we may have turbines lowered in the oceans and turned by the movement of tides as they come in and go out. The use of fuel cells holds promise as well.

Electricity is created by electrons in an atom being pushed from one atom to the next. The atom consists of three principal parts: the neutron, the proton, and the electron. The neutron and proton form the nucleus or center of the atom. The electrons orbit around the nucleus, much like the earth orbits around the sun. Each material has different numbers of electrons orbiting around the nucleus. Hydrogen has one electron, which means there is only one orbit around the nucleus. The first orbit has only two electrons in it. The next orbit has four electrons in it. These orbits continue until all of the electrons of an element are used. Copper has a total of 63.5 electrons in orbit around the nucleus. The outermost orbit contains only one-and-a-half electrons, which makes it easy to push them from one atom to the next.

The pushing of these electrons is measured in volts. The faster the pushing of these electrons, the higher the voltage. Some materials allow their electrons to be pushed more freely than others. Examples of these materials are gold, silver, copper, and aluminum. This ability to allow their electrons to be pushed from one atom to the next makes these materials good conductors. Other materials are more resistant to allowing their electrons to be pushed from one atom to the next. Examples of these materials are paper, glass, pure water, and plastic. This ability to resist the flow of electrons from one atom to the next makes these materials insulators. The outermost orbits of electrons in these materials are fuller and therefore resist the pushing of electrons from one atom to the next.

The electrons are pushed on the surface of the conductor rather than through it. Stranded conductors provide more surface area for the electrons to move than do solid conductors of the same size.

As mentioned earlier, 95% of the electrical power generated in this country is generated at 13,800 volts, so standardized equipment can be utilized, thereby keeping the costs down, as well as reducing handling problems. The voltage typically is transmitted from the turbines to a substation. At the substation, the 13,800 volts is stepped up via a transformer, typically to 138,000 volts. This is the beginning of the transmission of high voltage and the end of the generation phase.

At this point, it should be mentioned that although voltage is what pushes the electrons down the conductor, it is the amperage that represents the amount of work electricity does. An important point is that voltage and amperage are directly proportional, that is, volts × amperes equals watts. If you double the voltage, the amperage is cut in half, resulting in the same wattage. For example,

$$120 \text{ volts} \times 10 \text{ amperes} = 1200 \text{ watts}$$

$$240 \text{ volts} \times 5 \text{ amperes} = 1200 \text{ watts}$$

Bear this in mind as we walk through an electrical system from generation to home. A turbine is churning out 15 megawatts of power, at 13,800 volts. A bus duct of 2000 ampere rating takes the voltage to a transformer. The voltage is stepped up to 138,000 volts, but since the voltage is stepped up ten times, the amperage is only one-tenth, so the wattage is the same on both sides of the transformer.

The square root of three, or 1.732, used in the calculation below is common with three-phase calculations. This will be shown in future chapters.

$$13,800 \times 2000 \text{ amperes} \times 1.732 = 47,803,200 \text{ watts}$$

$$138,000 \times 200 \text{ amperes} \times 1.732 = 47,803,200 \text{ watts}$$

The wire size for the transmission line would only need to carry 200 amperes, which means that the wire size could be considerably smaller, which makes it more economical as well as practical. When the transmission line gets to its destination, a city in this example, it would be stepped down through a transformer to a more practical voltage of 12,470 volts. (This voltage varies with different power companies.) At this point, the transmission of the higher voltage ends and the distribution of the lower voltage begins. This phase distributes the voltage around the city to homes and businesses, both overhead and underground. Now there is enough power for over 150,000 homes. The voltage is distributed in several smaller 12 kilovolt lines around town. As it reaches a neighborhood, a typical transformer will step the voltage down from the distribution voltage of 12,470 volts to 115 or 230 volts.

Several factors were not considered in the above example. It should only serve as a simplified guide to the generation, transmission, and distribution of electrical power. Distance, line loss, transformer loss, and voltage drop were not considered. Also, transmission lines handle much larger amperage loads than the one given in this example.

chapter 1

DIRECT CURRENT CALCULATIONS

1-1   OHM'S LAW

Ohm's Law may be divided into four elements: watts (P), voltage (E), amperage (I), and ohms (R) (see Figure 1-1):

watts (P) = voltage × amperage ($EI = P$)

= ohms × amperage^2 ($RI^2 = P$)

= voltage^2 divided by ohms $\left(\dfrac{E^2}{R} = P\right)$

voltage (E) = ohms × amperage ($RI = E$)

= watts divided by ohms $\left(\dfrac{P}{R} = E\right)$

= square root of watts × ohms ($\sqrt{PR} = E$)

ohms (R) = Voltage divided by amperage $\left(\dfrac{E}{I} = R\right)$

= voltage squared divided by watts $\left(\dfrac{E^2}{P} = R\right)$

= watts divided by amperage^2 $\left(\dfrac{P}{I^2} = R\right)$

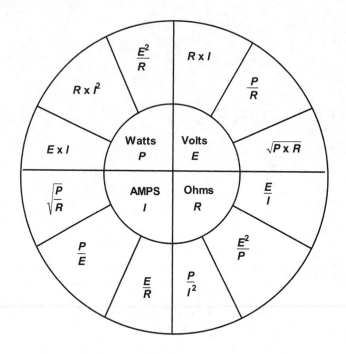

Figure 1-1

$$\text{amperage } (I) = \text{voltage divided by ohms} \left(\frac{E}{R} = I \right)$$

$$= \text{watts divided by ohms} \left(\frac{P}{R} = I \right)$$

$$= \text{square root of watts divided by ohms} \left(\sqrt{\frac{P}{R}} = I \right)$$

To show how these equations work, a simple circuit of 120 volts at 6 amperes with a resistance of 20 ohms is used. Watts in this circuit is $EI = P$ or $120 \times 6 = 720$ watts.

$$\text{To find watts} = EI \; 120 \times 6 = 720 \text{ watts}$$

$$= RI^2 = 20 \times 6^2 = 20 \times 36 = 720 \text{ watts}$$

$$= \frac{E^2}{R} = \frac{120^2}{20} = \frac{14400}{20} = 720 \text{ watts}$$

To find voltage $= RI = 20 \times 6 = 120$ volts

$$= \frac{P}{R} = \frac{720}{20} = 120 \text{ volts}$$

$$= \sqrt{PR} = \sqrt{720 \times 20} = \sqrt{14400} = 120 \text{ volts}$$

To find ohms $= \dfrac{E}{I} = \dfrac{120}{6} = 20$ ohms

$$= \frac{E^2}{R} = \frac{120^2}{20} = \frac{14400}{20} = 20 \text{ ohms}$$

$$= \frac{P}{I^2} = \frac{720}{6^2} = \frac{720}{36} = 20 \text{ ohms}$$

To find amperage $= \dfrac{E}{R} = \dfrac{120}{20} = 6$ amperes

$$= \frac{P}{E} = \frac{720}{120} = 6 \text{ amperes}$$

$$= \sqrt{\frac{P}{R}} = \sqrt{\frac{720}{20}} = \sqrt{36} = 6 \text{ amperes}$$

Much information can be calculated from these equations. For these equations to work, only two parts of the equation need to be known. For example, a 200 watt lighting fixture is on 120 volt circuit. How many amperes does it draw?

$$\frac{P}{E} = I \qquad \text{or} \qquad \frac{200}{120} = 1.67 \text{ amperes}$$

Exercise 1-1

1. Name the four elements of Ohm's Law.
2. A circuit has 40 ohms resistance and a 5 ampere load. What is the wattage? What is the voltage?
3. How may ohms are in a circuit that has 840 watts at 120 volts? What is the amperage?
4. A circuit has 1800 watts. What is the voltage if the circuit draws 15 amperes?

5. What is the wattage of a circuit that draws 5 amperes at a resistance of 24 ohms?

Answers to Exercise 1-1

1. Watts (P), voltage (E), amperage (I), and ohms (R).
2. Watts $= RI^2 = 40 \times 6^2 = 40 \times 36 = 1440$ watts. Volts $= R \times I = 40 \times 6 = 240$ volts.
3. To find resistance $= \dfrac{E^2}{P} \; \dfrac{120^2}{840} = \dfrac{14,400}{840} = 17.14$ ohms. To find amper-

   age $= \dfrac{E}{R} = I = \dfrac{120}{17.14} = 7$ amperes
4. To find the voltage first find ohms: $\dfrac{P}{I^2} = R$; $\dfrac{1800}{15^2} = \dfrac{1800}{225} = 8$ ohms; $RI = E$; $8 \times 15 = 120$ volts.
5. Watts $= RI^2$ $24 \times 5^2 = 24 \times 25 = 600$ watts.

1-2 SERIES CIRCUIT

The basic formula is stated as voltage is equal to resistance times amperage. The problem lies in applying the basic formula to circuits in the proper manner. First of all, determine the type of circuit that you have: parallel, series, or a combination of the two.

For the series circuit, the basic rule is the same: $E = I \times R$. Each resistor has its own voltage drop, its own amperage, and its own resistance. For instance, a resistor at 10 ohms has a voltage drop of 20 volts. Now apply ohms law to it:

$$E = I \times R, \qquad E = 20 \text{ volts}, \qquad I = ?$$
$$R = 10 \text{ ohms}$$
$$20 = I \times 10$$

To find I, divide R into E:

$$\frac{E}{R} = \frac{20}{10} = 2 \text{ amperes}$$

In a series circuit, the amperage across one resistor is the same across all resistors. The total amperage is the same as the amperage across each resistor:

$$I_T = I_1 = I_2 = I_3 \quad \text{etc.}$$

In the above example, $I = 2$ amperes. The circuit therefore has a total of 2 amperes and each resistor in that circuit has a 2 ampere load across it.

Voltages in a series circuit add together. The total voltage of a series circuit is the sum of the voltages across all the resistors:

$$E_T = E_1 + E_2 + E_3 \quad \text{etc.}$$

Resistances in a series circuit add together. The total resistance of a circuit is equal to the sum all the resistors added together:

$$R_T = R_1 + R_2 + R_3 \quad \text{etc.}$$

For example, consider three resistors in series having a value of 3 ohms, 2 ohms, and 5 ohms, as shown in Figure 1-2. The circuit has 2 amperes across it. The voltage drop is determined by $E = I \times R$:

$R_1 = 3$ ohms
$R_2 = 2$ ohms
$R_3 = 5$ ohms
$R_T = 10$ ohms (where R_T is the total resistance)

Multiply each resistor times the amperage to find each voltage drop:

$E_1 = 3$ ohms $\times$ 2 amperes $= 6$ volts
$E_2 = 2$ ohms $\times$ 2 amperes $= 4$ volts

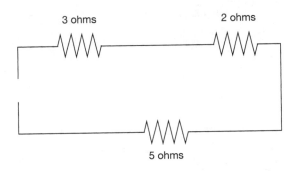

3 ohms   2 ohms

5 ohms

Figure 1-2

$E_3 = 5$ ohms $\times$ 2 amperes $= 10$ volts

$E_T = 20$ (where E_T is the total voltage drop)

Problem: Look at Figure 1-3 and find the following:

I_1 _____ E_1 _____ R_T _____
I_2 _____ E_2 _____ I_T _____
I_3 _____ E_3 _____ E_T _____

$$R_T = R_1 + R_2 + R_3 = 10 + 5 + 5 = 20$$

$$R_T = 20 \text{ ohms}$$

$$E_T = 100 \text{ volts}$$

To find I_T, divide E_T by R_T:

$$\frac{100}{20} = 5 \text{ amperes}$$

$I_1 = I_2 = I_3 = I_T$
$I_1 = 5, \qquad I_2 = 5, \qquad I_3 = 5, \qquad I_T = 5$

$E_1 = I_1 \times R_1$	or	5×10	$= 50$ volts
$E_2 = I_2 \times R_2$	or	5×5	$= 25$ volts
$E_3 = I_3 \times R_3$	or	5×5	$= 25$ volts
$E_T = I_T \times R_T$	or	5×10	$= 100$ volts

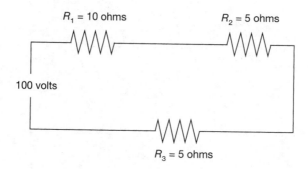

$R_1 = 10$ ohms $R_2 = 5$ ohms

100 volts

$R_3 = 5$ ohms

Figure 1-3

Exercise 1-2

1. Three resistors in series have a value of 4 ohms, 2 ohms, and 1 ohm. The voltage drop across the 2 ohm resistor is 10 volts. What is the total resistance, total amperage, and total voltage of the circuit.

2. Four resistors have a total voltage of 100 volts. The second resistor has a voltage of 20 volts at 2 amperes. Resistor 3 has a voltage drop of 30 volts. Resistor four has a voltage drop of 10 volts. What is the value of each resistor in this circuit?

3. Three equal resistors are connected in series at 5 ohms each. The voltage at each resistor is 25 volts. Give the total resistance of the circuit, the total voltage of the circuit, and the total amperage of the circuit.

4. A circuit with a resistance of 60 ohms has an ampacity of 2 amperes. What is the voltage?

5. A circuit has a total voltage of 120 volts at 40 ohms resistance. What is the circuit ampacity?

Answers to Exercise 1-2

1. $R_T = R_1 + R_2 + R_3 = 4 + 2 + 1 = 7$ ohms
$I_T = I_1 = I_2 = I_3$

$$\frac{E_2}{R_2} = \frac{10}{2} = 5 \text{ amperes}$$

$I_1 = I_2 = I_3 = I_T = 5 = 5 = 5$ amperes
$E_1 = R_1 I_1 = 4 \times 5 = 20$ volts
$E_2 = R_2 I_2 = 2 \times 5 = 10$
$E_3 = R_3 I_3 = 1 \times 5 = 5$
$E_T = E_1 + E_2 + E_3 = 20 + 10 + 5 = 35$ volts

2. $I_T = I_1 = I_2 = I_3 = I_4$
$E_T = E_1 + E_2 + E_3 + E_4$
$100 - E_2 - E_3 - E_4 = E_1$
$100 - 20 - 30 - 10 = E_1 = 40$ volts

$$\frac{E_1}{I_1} = R_1$$

$$\frac{40}{2} = 20 \text{ ohms}$$

$$\frac{E_2}{I_2} = R_2 = \frac{20}{2} = 10 \text{ ohms}$$

$$\frac{E_3}{I_3} = R_3 = \frac{30}{2} = 15 \text{ ohms}$$

$$\frac{E_4}{I_4} = R_4 = \frac{10}{2} = 5 \text{ ohms}$$

$R_1 = 20$ ohms
$R_2 = 10$ ohms
$R_3 = 15$ ohms
$R_4 = 5$ ohms

3. $R_T = R_1 + R_2 + R_2$
$5 + 5 + 5 = 15$ ohms
$E_T = E_1 + E_2 + E_3$
$25 + 25 + 25 = 75$ volts
$I_T = I_1 = I_2 = I_3$
$$I_T = \frac{E_T}{R_T} = \frac{75}{15} = 5 \text{ amperes}$$

4. $E_T = R \times I \ 60 \times 2 = 120$ volts

5. $\dfrac{E}{R} = I$

$$\frac{120}{40} = 3 \text{ amperes}$$

1-3 PARALLEL CIRCUITS

The same general rule applies: $E = I \times R$. However, a slightly more complicated method is applied. Voltage is the same throughout the circuit and amperage is equal to the sum of the amperage across each resistor:

$$I_T = I_1 + I_2 + I_3 \text{ etc.}$$

Resistance is calculated differently depending on the number and size of the resistors. When two resistors of unequal value are used in parallel, the values of the resistors are multiplied together, then divided by their sum:

$$\frac{R_1 \times R_2}{R_1 + R_2}$$

Two resistors in parallel have a value of 2 ohms and 4 ohms. To find the value of these resistors in parallel, use the above formula.

$$R_1 = 2 \text{ ohms} \qquad R_2 = 4 \text{ ohms}$$

$$R_T = \frac{2 \times 4}{2 + 4} = \frac{8}{6} = 1.33$$

It is helpful to note that the total resistance of a parallel circuit is less than the smallest resistor. If two or more resistors are in parallel and are equal, then you divide the value of the resistor by the number of resistors:

$$\frac{R}{N} = R_T$$

For instance three resistors are in parallel, each having a value of 6 ohms:

$$R_T = \frac{6}{3} = 2 \text{ ohms}$$

If more than two resistors of unequal value are used, the following would apply:

$$\frac{1}{R_1} + \frac{1}{R_2} + \frac{1}{R_3} = \frac{1}{R_T}$$

Three resistors in parallel have a value of 2 ohms, 4 ohms, and 6 ohms. What is the total resistance?

$$\frac{1}{2} + \frac{1}{4} + \frac{1}{6} = \frac{1}{R_T}$$

First, you must find a common denominator when adding fractions—a number that 2, 4, and 6 will go into equally. The smallest number that 2, 4, and 6 will go into would be 12.

Now the fractions can be turned into portions of 12:

$$\frac{1}{2} = \frac{6}{12}, \frac{1}{4} = \frac{3}{12}, \frac{1}{6} = \frac{2}{12}$$

Next you line up the new fractions and add them together:

$$\frac{6}{12} + \frac{3}{12} + \frac{2}{12} = \frac{11}{12}$$

To find R_T, invert the equation or turn it upside down, from

$$\frac{11}{12} = \frac{1}{R_T}$$

to

$$\frac{12}{11} = R_T$$

Therefore,

$$R_T = \frac{12}{11}$$

or $R_T = 1.09$ ohms.

Once the resistance value of a parallel circuit is obtained, the rest of Ohm's law can be applied: $E = I \times R$.

To find the amperage of a parallel resistor within a parallel circuit, find the total resistance and total voltage of the circuit. Divide the voltage by the resistance to find the total amperes. Go back and divide the voltage by each resistor to find the amperage of each. Add the amperes of each resistor and the total will equal I_T.

Example. A 5 ohm and 10 ohm resistor are in parallel. The circuit has a total of 100 volts. Find the amperage across each resistor and the total amperage of the circuit.

First, divide the voltage across the first resistor by the value of the resistor:

$$\frac{100}{5} = 20 \text{ amperes} \qquad I_1 = 20 \text{ amperes}$$

Now do R_2 the same way:

$$\frac{100}{10} = 10 \text{ amperes} \qquad I_2 = 10 \text{ amperes}$$

Now add I_1 and I_2 together:

$$20 + 10 = 30$$

The total resistance of the circuit is found by using the formula for two unequal resistors:

$$\frac{R_1 \times R_2}{R_1 + R_2} = E = \frac{50}{15} = 3.34 \text{ ohms}$$

Now apply Ohm's law:

$$E = R \times I$$

$$I = \frac{E}{R}$$

$$\frac{100}{3.34} = 29.9 \text{ or } 30 \text{ amperes} = I_T$$

Exercise 1-3

1. What is the total resistance of three resistors in parallel with an equal value of 18 ohms?

2. A circuit has a voltage of 200 with two resistors of 10 ohms and 15 ohms in parallel. What is the amperage across each resistor?

3. Two resistors in parallel of unequal value have the following amperages: $I_1 = 5$ amperes and $I_2 = 10$ amperes. What is the value of each resistor if 100 volts is applied across this circuit? What is the total resistance of the circuit?

4. A circuit with 150 volts has two resistors in parallel: $R_1 = 3$ ohms and $R_2 = 6$ ohms. Find the amperage across each resistor and the total resistance.

5. Two groups of resistors are in parallel. The first group has two resistors of 4 ohms each in parallel. The second group of parallel resistors has an 8 ohm resistor in parallel with a 12 ohm resistor. Find the total resistance and the amperage across each resistor if 160 volts is applied to the circuit.

Answers to Exercise 1-3

1. $\dfrac{R}{N} = R_T$

   $\dfrac{18}{3} = 6 \text{ ohms}$

2. $I_1 = \dfrac{E}{R_1} = \dfrac{200}{10} = 20$ amperes

$I_2 = \dfrac{E}{R_2} = \dfrac{200}{15} = 13.33$ amperes

3. $\dfrac{E}{I_1} = R_1 = \dfrac{100}{5} = 20$ ohms

$\dfrac{E}{I_2} = R_2 = \dfrac{100}{10} = 10$ ohms

$R_T = \dfrac{R_1 \times R_2}{R_1 + R_2}$

$\dfrac{20 \times 10}{20 + 10} = \dfrac{200}{30} = 6.67$ ohms

4. $\dfrac{E}{R_1} = I_1 = \dfrac{150}{3} = 50$ amperes

$\dfrac{E}{R_2} = I_2 = \dfrac{150}{6} = 25$ amperes

$R_T = \dfrac{R_1 \times R_2}{R_1 + R_2} = \dfrac{3 \times 6}{3 + 6} = \dfrac{18}{9} = 2$ ohms

5. $R_A = \dfrac{R}{N} = \dfrac{4}{2} = 2$ ohms

$R_B = \dfrac{R_3 \times R_4}{R_3 + R_4} = \dfrac{8 \times 12}{8 + 12} = \dfrac{96}{20} = 4.8$ ohms

$R_T = R_A + R_B = 2 + 4.8 = 6.8$ ohms

$I_T = \dfrac{160}{6.8} = 23.5$ amperes

Find E_A and E_B:
$E_A = R_A \times I_T = 2 \times 23.5 = 47$ volts
$E_B = R_B \times I_T = 4.8 \times 23.5 = 112.8$ volts

$I_1 = \dfrac{E_A}{R_1} = \dfrac{47}{4} = 11.75$ amperes

$I_2 = \dfrac{E_A}{R_2} = \dfrac{47}{4} = 11.75$ amperes

$$I_3 = \frac{E_B}{R_3} = \frac{112.8}{8} = 14.1 \text{ amperes}$$

$$I_4 = \frac{E_B}{R_4} = \frac{112.8}{12} = 9.4 \text{ amperes}$$

1-4 COMBINATION CIRCUITS

This is by far the most confusing calculation. A combination circuit can be made up of resistors in series with resistors in parallel with resistors in series and parallel. In Figure 1-4, Resistors R_1 and R_8 are in series. Resistors R_2 and R_3 are in series with each other but in parallel with R_4. Resistor R_6 is in series with resistor R_7 but added together becomes parallel with R_5. First add $R_2 + R_3$, which is in series: $1 + 1 = 2$. Let us call this resistor R_A. R_A is parallel with R_4 (see Figure 1-5). The equivalent resistance of re-

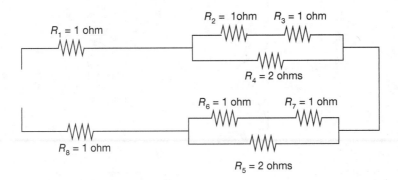

Figure 1-4

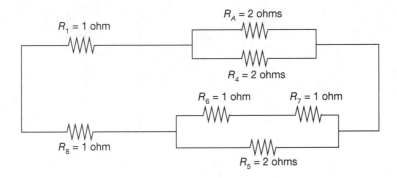

Figure 1-5

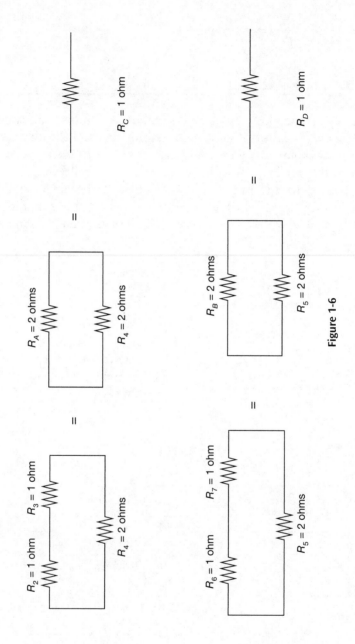

Figure 1-6

18

sistors R_2, R_3, and R_4 is converted by first adding the resistors in series—$R_2 + R_3$—resulting in R_A, then using the formula below to convert the parallel resistors R_A and R_4 to an equivalent series resistor:

$$R_A = 2 \text{ ohms} \qquad R_4 = 2 \text{ ohms}$$

Apply

$$\frac{R}{N} = \frac{2}{2} = 1 \text{ ohm}$$

Do the same with R_6 in series with R_7:

$$1 + 1 = 2$$

Call this R_B. Figure 1-6 shows the resistors being combined and converted to the equivalent resistance so that they can be added together, much like converting fractions of different denominators to a common denominator so that they can be added together.

Redraw the new circuit with the new equivalent resistors R_C and R_D along with the resistors that were orignally in series, as in Figure 1-7:

$$R_1 + R_C + R_D + R_8 = 1 + 1 + 1 + 1 = 4 \text{ ohms}$$

$$R_T = 4 \text{ ohms}$$

$$I_T = \frac{E_T}{R_T} = \frac{240}{4} = 60$$

$$I_T = 60 \text{ amperes}$$

Since amperage is the same throughout a series circuit, each resistor or group of resistors will have 60 amperes across it, so $I_1 = 60$, $I_C = 60$, $I_D = 60$, and $I_8 = 60$. Multiply each I by R:

$$60 \times 1 = 60 \text{ volts} = E_1$$
$$60 \times 1 = 60 \text{ volts} = E_C$$
$$60 \times 1 = 60 \text{ volts} = E_D$$
$$60 \times 1 = 60 \text{ volts} = E_8$$
$$\overline{240 \text{ volts} = E_T}$$

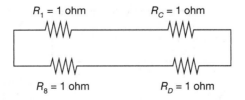

$R_1 = 1$ ohm $\qquad R_C = 1$ ohm

$R_8 = 1$ ohm $\qquad R_D = 1$ ohm

Figure 1-7

Now look at R_C. Remember that R_C is made up of resistors in parallel and in series. The whole group is R_C, but is broken down to R_A and R_4 (see Figure 1-8). The total voltage of $R_C = 60$ volts, the total amperage for $R_4 = 60$ amperes.

The voltage across R_A and R_4 is the same: 60 volts. The amperage for R_A and R_4 can be found by dividing E_C by R_A:

$$\frac{E_C}{R_4} = \frac{60}{2} = 30 \text{ amperes}$$

Divide E_C by R to find the amperage:

$$\frac{E_C}{R_C} = 30 \text{ amperes}$$

$$I_A = 30 \text{ amperes} \qquad I_4 = 30 \text{ amperes}$$

Amperage in parallel add together, so $I_A + I_4 = I_C = 30 + 30 = 60$ amperes. R_A is equivalent to R_2 and R_3 added together. Now, the total of R_A equals 2 ohms, but R_2 and R_3 have a value of 1 ohm each and the circuit looks like the one in Figure 1-9.

Each resistor times the circuit amperage equals the voltage drop across each resistor:

$$R_2 \times I = E_2 = 1 \times 30 = 30$$
$$R_3 \times I = E_3 = 1 \times 30 = 30$$
$$E_2 + E_3 = E_A = 30 + 30 = 60 \text{ volts}$$

Now do the same thing to Group D:

$$E_D = 60 \text{ volts}$$

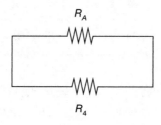

R_A

R_4

Figure 1-8

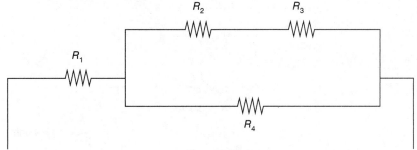

Figure 1-9

$$I_D = 60 \text{ amperes}$$

$$R_D = R_B \text{ paralleled with } R_5$$

$$R_B = 2 \text{ ohms} \qquad R_5 = 2 \text{ ohms}$$

$$I_B = \frac{60}{2} = 30 \text{ amperes} \qquad I_5 = \frac{60}{2} = 30 \text{ amperes}$$

$$I_B + I_5 = I_D = 30 + 30 = 60 \text{ amperes}$$

$$R_B = R_6 + R_7$$

$$I_B = 30 \text{ amperes}$$

$$E_B = 60 \text{ volts}$$

$$R_6 = 1 \text{ ohm} \qquad I_6 = 30 \text{ amperes}$$

$$R_7 = 1 \text{ ohm} \qquad I_7 = 30 \text{ amperes}$$

$$R_6 \times I_6 = E_6 = 1 \times 30 = 30 \text{ volts}$$

$$R_7 \times I_7 = E_7 = 1 \times 30 = 30 \text{ volts}$$

$$E_6 + E_7 = E_B = 30 + 30 = 60 \text{ volts}$$

The important thing is to break the circuit down into equivalent resistors and work each unit out.

Find the voltage drop across each resistor in Figure 1-10, where $E_T = 120$ volts. First find R_T. All resistors are in series except R_2 and R_3, which are in parallel. Convert R_2 and R_3 to an equivalent resistor,

$$R_A = \frac{R_2 \times R_3}{R_2 + R_3} = \frac{3 \times 6}{3 + 6} = \frac{18}{9} = 2$$

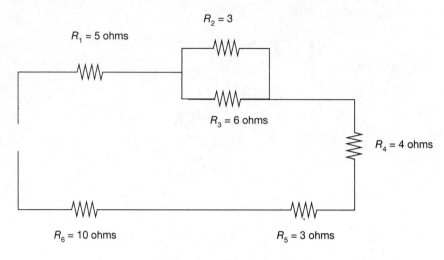

Figure 1-10

Now add all the resistors together.

$$R_T = R_1 + R_A + R_4 + R_5 + R_6 \quad \text{or} \quad 5 + 2 + 4 + 2 + 7 = 20 \text{ ohms}$$

Now multiply each resistor by the amperage:

$$
\begin{aligned}
E_1 &= I_T \times R_1 = 5 \times 6 = 30 \text{ volts} \\
E_A &= I_T \times R_A = 2 \times 6 = 12 \text{ volts} \\
E_4 &= I_T \times R_4 = 4 \times 6 = 24 \text{ volts} \\
E_5 &= I_T \times R_5 = 2 \times 6 = 12 \text{ volts} \\
E_6 &= I_T \times R_6 = 7 \times 6 = 42 \text{ volts} \\
\hline
&\qquad\qquad\qquad\qquad\quad 120 \text{ volts}
\end{aligned}
$$

Since $E_A = E_2 = E_3$, in a parallel circuit 12 volts is across E_2 and E_3. First find R_T. All resistors are in series.

Exercise 1-4

The following problems are based on Figure 1-11.

1. Find R_T
2. Find I_T

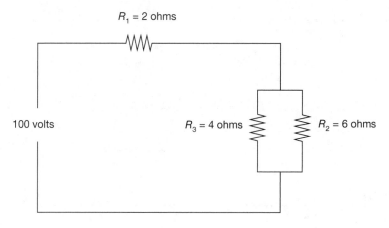

Figure 1-11

3. Find E_1, E_2, and E_3
4. Find I_1, I_2, and I_3
5. Find P_T

Answers to Exercise 1-4

1. To find R_T combine R_2 and R_3:

$$R_2 + R_3 = \frac{R_2 \times R_3}{R_2 + R_3} = \frac{4 \times 6}{4 + 6} = \frac{24}{10} = 2.4 \text{ ohms}$$

Next add R_1:

$$2.4 + 2 = 4.4 \text{ ohms}$$

2. To find I_T, use

$$\frac{E_T}{R_T} = \frac{100}{4.4} = 22.7 \text{ amperes}$$

3. To find E_1, use $R_1 \times I = 2 \times 22.72 = 45.44$ volts. To find E_2, find the voltage drop across the equivalent resistor; multiply $I_T \times 2.4 = 22.7 \times 2.4 = 54.48$ volts. Since voltage is the same across parallel resistors, $E_2 = E_3$ and both = 54.48 volts.

4. $I_1 = I_T = 22.7$ amperes

$$I_2 = \frac{E_2}{R_2} = \frac{54.48}{6} = 9.08 \text{ amperes}$$

$$I_3 = \frac{E_3}{R_3} = \frac{54.48}{4} = 13.62$$

5. To find P_T, use $E_T \times I_T = 100 \times 22.72 = 2272$ watts.

CHAPTER 1 TEST*

1. Two resistors of 2 ohms each are in series. If the circuit has a total of 120 volts, what is the total amperage?

2. If two resistors of 4 ohms each are in parallel and the total voltage is 120 volts, what is the total amperage?

3. If a circuit has a total of 1800 watts and amperage of 10 amperes, what is the voltage?

4. 3 resistors in parallel: $R_1 = 6$ ohms, $R_2 = 4$ ohms, and $R_3 = 3$ ohms. What is the total resistance of the circuit?

5. If a circuit has 400 watts and a voltage of 100 volts, what is the total resistance of the circuit?

6. Three resistors each have a value of 6 ohms and are in parallel. What is the equivalent resistance?

Problems 7 through 11 are based on Figure 1-12.

7. In Figure 1-11, what is R_T?

8. What is the total amperage of the circuit?

9. What is the voltage drop across R_2?

10. What is the equivalent resistance of R_5, R_6, and R_7?

11. What is the amperage across the equivalent resistance of R_5, R_6, and R_7?

12. What is the equation for finding the total resistance in parallel of three unlike resistors?

13. In a series circuit of 120 volts and three resistors of $R_1 = 2$, $R_2 = 4$, and $R_3 = 6$, what is the amperage across each resistor? What is the total amperage of the circuit?

*Answers to tests follow Chapter 11.

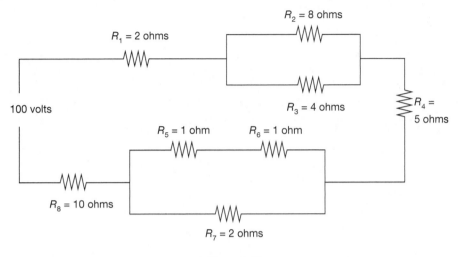

Figure 1-12

14. A 50 volt circuit has a resistance of 5 ohms, what is the watts?

15. A series circuit of three resistors of 3 ohms, 6 ohms, and 8 ohms has a total of 5 amperes across it. What is the voltage drop across each resistor? What is the total voltage of the circuit?

16. In a parallel circuit, two resistors of 10 ohms each have an output of 500 watts. What is the voltage of this circuit?

17. What is the total amperage of Problem 16?

18. In a series circuit, $E_T = E_1 + E_2 + E_3$. Is this true or false?

19. What is the amperage for two resistors of 5 ohms and 10 ohms in series if the voltage drop across the 5 ohm resistor is 50 volts?

20. What is the total voltage of the circuit in Problem 19?

21. Two resistors are in parallel. One is 10 ohms and the second is 20 ohms on a 240 volt circuit. What is the amperage across each resistor?

22. What is the total amperage in Problem 21?

23. If a resistor has 4 amperes across it with a voltage drop of 40 volts, what is the resistance?

24. What is the voltage of a circuit if two resistors are in parallel, each having a value of 6 ohms and drawing 8 amperes?

25. Voltage is the product of amperage times resistance. True or false?

chapter 2

ALTERNATING CURRENT CALCULATIONS

2-1 SINGLE PHASE

Single phase is two wires connected to a load. This can be phase to ground or phase to phase. For example, a 110 volt load is a phase-to-ground load. A 200 volt load is a phase-to-phase load. In order to be single phase, the 220 volt load must consist of only two wires (not counting the equipment ground) with a potential difference of 220 volts between them (see Figure 2-1).

To find the power for the above examples, lets make each load 10 amperes.

$$P = I \times E$$

Power is also watts:

$$110 \text{ volts} \times 10 \text{ amperes} = 1100 \text{ watts}$$

$$220 \text{ volts} \times 10 \text{ amperes} = 2200 \text{ watts}$$

Single phase is used primarily for lighting and small loads such as heat, air conditioning, fan motors, and other small motors.

There are various combinations of power that will deliver single-phase loads. As stated, small loads can be utilized on single phase either phase to phase or phase to ground. One of the most common systems is a single-phase transformer supplying 120 volts to ground from either phase, or 240 volts between the two phases (see Figure 2-2).

This is common for residences and small buildings. As the load increases, it is usually more favorable to go to a three-phase system. One

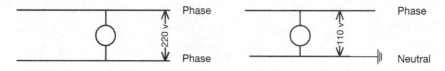

Figure 2-1

such system is a three-phase, four-wire, solidly grounded 120/208 volt system. Sometimes, this is referred to a three-phase 120 volt system. The fourth wire is the neutral or grounded conductor. Each ungrounded conductor has a voltage of 120 volts between it and the neutral. This is an excellent system for large amounts of lighting and other phase-to-ground loads. This type of loading can be closely balanced on all three ungrounded conductors. Also, it has three-phase capabilities. This system is very popular as a separately derived system. A separately derived system is nothing more than one or more transformers, usually inside a building. In some areas, the power company will give a lower rate to a customer for utilizing one voltage at the service entrance. For instance, it is more practical to use 480/277 volts at the service entrance in larger buildings. The 480 volt three phase can be used for the larger motors and the 277 volt can be used for much of the lighting. However, there undoubtedly will be receptacles and possibly some 120 volt incandescent lighting as well as some small single-phase and three-phase loads that would have to be served by a separately derived system, namely 120/208 volts.

If this building had 500 100-watt lights that would be 500 × 100 = 50,000 watts. If three phases to ground of 277 volts were used,

$$\frac{50,000}{3 \times 277} = 60.16 \text{ amperes per phase and neutral}$$

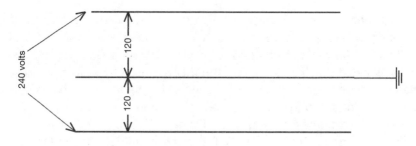

Figure 2-2

However, if 120/208 volts were used instead,

$$\frac{50,000}{3 \times 120} = 138.8 \text{ amperes per phase and neutral}$$

It can easily be seen that the 277 volt lighting has less than half the amperage that the 120 volt lighting has. The same holds true for motors and other loads: the higher the voltage, the lower the amperage. Since wire size is determined by amperage, and conduit size by wire size, the separately derived system will often be justified.

An interesting thing that should be mentioned happens on a separately derived system. In this case the 500 100-watt lights will be used, and let us use a 120 amp motor load at 480 volts, three phase, and 100 amps for receptacles and other 120 volt loads (see Figure 2-3).

Although there are neutrals on the 480 volt side and on the 208 volt side, they have no relation with each other or the three phases of the 480 that were used to produce the three-phase four-wire 120/208 with a neutral. Note that the neutral is absorbed through the coils of the transformer and is not present on the 480 volt side of the transformer.

Another popular system, particularly for medium-sized or small buildings with three-phase requirements, is the delta high-leg, three-phase, four-wire 240–208–120 volt system. This system, however, only provides 120 volts from two phases to ground. The third phase yields 208 volts to ground and, therefore, cannot be used for lighting and receptacles. In or-

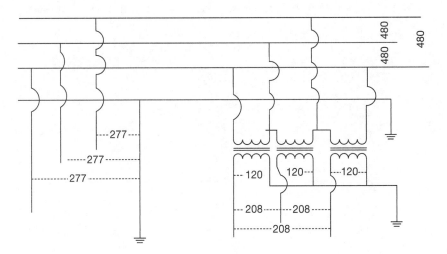

Figure 2-3

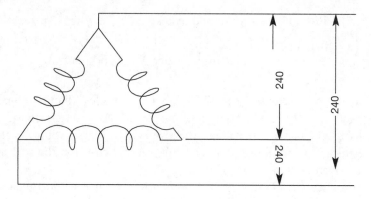

Figure 2-4

der to understand what happens in this case, a brief explanation of the delta configuration is necessary. This and other transformer hookups will be covered in more detail later.

Each side of the triangle represents one transformer (see Figure 2-4). The apex is the angle formed by the ends of two transformer windings tied together. This yields 240 volts between any two phases or all three phases. This is a straight delta.

In order to achieve lighting from this, one of the transformers must have its midpoint grounded. This does not in any way effect the three phase or single phase to phase loads. 240 volts are obtained between points A and B, A and C, and B and C, and 240 volts are retained between A, B, and C (see Figure 2-5). However, the midpoint ground splits the

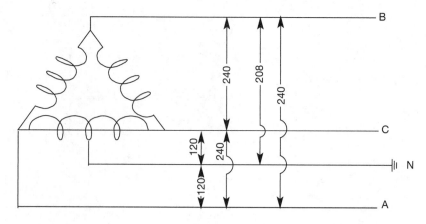

Figure 2-5

voltage in half between points N and A and N and C. This yields 120 volts at both locations. This provides lighting, receptacle, and other phases to ground-load capacities. However, a higher voltage to ground exists between N and B. The reason for this higher voltage is that a line between point B and N is longer.

The voltage between B and N can be calculated by the Pythagorean theorem since the line between B and N forms a right triangle. Hence, in calculating loads on this system for 120 volt loads, the formula is in watts/(2 × 120) or

$$\frac{\text{watts}}{240} = \text{amperes}$$

This would yield the amperage on both lighting phases to ground. For example,

$$\frac{3600}{240} \text{ watts of receptacles} = 15 \text{ amperes}$$

The *National Electrical Code*® is very clear on the marking and position of the higher phase to ground voltage. Article 384-3e of the *Code* says that the phase having the higher voltage to ground shall be marked. Article 215-8 states that the conductor having the higher voltage to ground shall be identified by having an outer finish that is orange in color or by tagging or other effective means. Article 334-3f states that the phase arrangement on three-phase buses shall be A, B, and C from front to back, top to bottom, or left to right as viewed from the front of the switchboard or panel board. Further, it states that the conductor having the higher voltage to ground shall be the B phase.

2-2   THREE-PHASE POWER

Three-phase power is extensively used in many applications and is more efficient than single-phase power. Larger motors are usually three phase, as are heating loads. Three-phase calculations involve the square root of 3 ($\sqrt{3}$):

$$E \times \sqrt{3} \times I = \text{watts}$$

$$\frac{\text{watts}}{E \times \sqrt{3}} = \text{amperes}$$

In many calculations, it is necessary to be able to convert from wattage to amperage and from amperage to wattage. For instance, a building may have its lighting in watts, motors in horsepower, heating in kilowatts, and appliances or equipment in amperage. All of these must be added together, and since amperage, wattage, and horsepower cannot be added together, they must be converted to identical terms in much the same way that fractions are. For example, combine the following:

60 kW heating, three phase, 480 volt

50,000 watts of lighting, 277 volts

62 ampere motor load three phase 480 volts

Convert to amperes:

$$60 \text{ kW} = \frac{60,000}{480 \times \sqrt{3}} \text{ watts} = 72.2 \text{ amperes}$$

$$\frac{50,000}{3 \times 277} \text{ watts} = 60.16$$

$$62 \text{ amperes} = \frac{62}{194.36}$$

Convert to watts:

$$60 \text{ kW} = 60 \times 1000 = 60,000 \text{ watts}$$

$$50 \text{ kW} = 50 \times 1000 = 50,000$$

$$62 \text{ amps} = 62 \times 480 \times \sqrt{3} = \frac{51522}{161,522}$$

Now convert final wattage to amperage:

$$\frac{161522}{480 \times \sqrt{3}} = 194.37 \text{ amperes}$$

Notice should be taken that the lighting was 277 volts yet the wattage was added directly to the 480 volt wattage and the sum was divided by 480 × √3. This is because 480 is the phase-to-phase voltage and 277 is the phase-to-ground voltage of the same system. Earlier, in discussing lighting, we used watts = amperes = 3 × 277 = 831:

$$480\sqrt{3} = 480 \times 1.732 = 831$$

The same holds true for 208/120:

$$\text{watts} = \text{amperes } 3 \times 120 = 360 = 208 \times \sqrt{3}$$

$$208 \times \sqrt{3} = 208 \times 1.732 = 360$$

However, note that this is true for solidly grounded wye-shaped systems, but not for the high-leg delta that was covered earlier. In the case of the high-leg delta, remember that the lighting (120 volt) load can only be used on two phases to ground. Therefore, all 120 volt loads must be calculated separately from three-phase or even single phase-to-phase loads. These can be calculated separately and the resulting amperage can be put on phases A, C, and the neutral. The three-phase load can be calculated as any three-phase load: voltage × $\sqrt{3}$ × amperage = watts.

For example, combine the following:

20 kW lighting (120 volt)

100 amperes of three-phase 240 volt

To convert to amperes, first make horizontal columns for each phase (lighting, three-phase load, and total), and vertical columns for the division of the loads (A, B, C, and N) (see Table 2-1). Next calculate each type of load. Note that similar loads may be added together, such as all 120 volt loads together and all 240 volt three-phase loads together:

$$\frac{20 \times 1000}{240} = 83.33 \text{ amperes}$$

other load = 100 amperes per phase

Table 2-1

	A	B	C	N
Lighting	83.3		83.3	83.3
Three-phase load	100	100	100	
Totals	183.3	100	183.3	83.3

Now fill in the table. Remember that the 120 volt load goes on A, C, and N. This yields the amperage on each conductor. The neutral will be looked at in more detail later in this chapter.

In the case of single phase, 240 volt loads can yield different results when calculating. The 240 volt single-phase load can be balanced between all three ungrounded phases. Remember that a 240 volt single-phase load is connected between any two phases. An approach to the problem from one direction would be to add the wattage together and apply it across all three phases. However, if the design splits the load coming into a three-phase panel, then going to a single-phase panel, the load would be equally connected between the A and C phases.

In the case of a branch circuit or feeder supplying single-phase loads of phase to phase, the neutral is not involved. Neither are neutral calculations on three-phase loads. The neutral need only concern itself with phase-to-neutral connections. In a delta system, to get the amperage of the neutral, divide the wattage by 240 volts, as only two phases to ground can be used for phase-to-neutral connections. If you divide the wattage by 120 volts, that will only yield the amperage on one phase. Divide this again by two and the two 120 volt phases are balanced. The results are for each phase and neutral. On a solidly grounded wye-shaped system, the phase-to-neutral voltages are equal between any phase to ground. A typical example is a 208/120 system. Each phase to ground is 120 volts; phase to phase or three ungrounded phase are 208 volts. On wye-shaped systems, the phase-to-phase voltage is 1.732 times the phase-to-ground voltage, or the phase-to-ground voltage is 1.732 divided into the phase-to-phase voltage:

$$120 \times 1.732 = 207.8 \text{ volts}$$

$$\frac{208}{1.732} = 120 \text{ volts}$$

The neutral load would be sized according to the maximum unbalance, so the phase with the largest amperage would be sized the same as the neutral. However, to determine the load on the neutral at a given time, the following formula can be used:

$$\sqrt{A^2 + B^2 + C^2 - AB - BC - AC}$$

For instance, the following loads were observed on a solidly grounded wye-shaped system (see Figure 2-6):

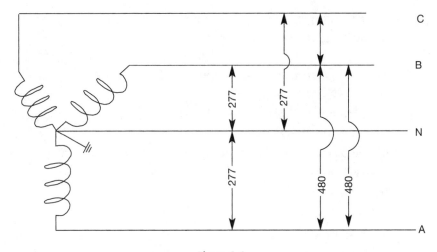

Figure 2-6

Phase A = 50 amperes

Phase B = 100 amperes

Phase C = 25 amperes

Neutral current = ?

$$\sqrt{50 \times 100 \times 25 - 50 \times 100 - 100 \times 25 - 50 \times 25} =$$

$$\sqrt{2500 + 10000 + 625 - 5000 - 2500 - 1250} = 66.14 \text{ amperes}$$

2-3 POWER FACTOR

Power plant "A" has power factor of 80% and volt amperes of 600 (see Figure 2-7). Power plant B has power factor of 90% and 400 watts. Combine

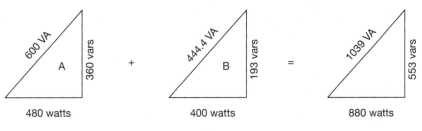

Figure 2-7

power plant A and power plant B. Find the new volt amperes, watts, vars (volt amperes reactive), and power factor (pf).

To find watts when volt amperes and the pf are known, multiply volt amperes × pf: 600 × 0.8 = 480

To find volt amperes when watts and the pf are known, divide the watts by the pf: 400/0.9 = 444.4

To find the pf when the volt amperes and watts are known, divide volt amperes into watts:

$600^2 - 480^2 = 360^2$

$444.4^2 - 400^2 = 193^2$

watts = 480 + 400 = 880

vars = 360 + 193 = 553 A = $880^2 + 553^2 = 1039^2$

$444.4^2 - 400^2 = 193^2$

New power factor = $\dfrac{880}{1039}$ = 84.6%

Sometimes it is necessary to improve the power factor. Ways to improve the pf include use of synchronous motors or capacitors. Synchronous motors and capacitors operate by adding leading vars to a circuit. As discussed later, leading vars subtract from lagging vars. When the vars are reduced, the phase angle is reduced. This means it takes less apparent power supplied to the load of true power.

A power plant operates at 60% power factor with 800 watts of power. Find how many leading vars are needed to bring the power factor up to 90% (see Figure 2-8):

1066 – 387.25 = 678.25 leading vars needed

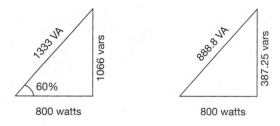

Figure 2-8

The power triangle is the last of three triangles explaining electrical relationships. Let us examine the three triangles and their relationship to each other.

Power

In order to calculate loads, a basic understanding of power is essential. The main unit of power that we will be concerned with is the watt. Calculations will frequently be converted to and from watts. One watt of power is one volt times one ampere. Watts are referred to as true power—the actual power being used by a circuit. Another expression of power is volt amperes. This is referred to as apparent power. It is the power being supplied to a circuit. Watts and volt amperes form a phase angle relationship (see Figure 2-9). The angle formed can be expressed as a percentage by dividing watts by volt amperes (see Figure 2-10).

The percentage of phase angle is called the power factor or pf. The power factor is also the cosine of the angle. If a power plant has a pf of 80%, the phase angle can be found by looking up 0.80 in the Table of Natural Trigonometric Functions (see page xxx) under the cosine column. This shows that the 70% pf is a 45.57 degree phase angle. If the 80% power factor was needed the phase angle would change to 0.80 = cosine of angle, or 36 degree phase angle. To improve the phase angle from 45 degrees to 36 degrees, the tangent of the smaller angle is subtracted from the tangent of the larger angle, then multiplied by the kilowatts. If the kilowatts is 800 kW, what is the kvars needed to correct the phase angle?

Cosine angle 1 = 0.7 or 45.57 degrees   Tangent angle 1 = 1.020

Cosine angle 2 = 0.8 or 36 degrees   Tangent angle 2 = (0.7265/
                                          0.2935) × 800 = 234.8 kvars

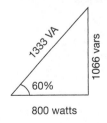
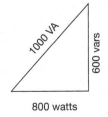

Figure 2-9

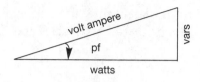

Figure 2-10

The loss of a circuit is expressed in vars (volt ampere reactive). This can be added to the phase angle to form a right triangle.

This completes the power triangle. Several calculations can be made from this triangle. In calculating relationships between the three sides of the triangle, the Pythagorean theorem for right triangles can be used:

$$A^2 + B^2 = C^2$$

where side A = watts, side B = vars, and side C = volt amperes.

If we have 500 volt amperes at 80% power factor, find watts and vars:

$$500 \times 0.80 = 400 \text{ watts}$$

$$\sqrt{500^2 - 400^2 = B^2}$$

$$\sqrt{2500 - 1600 = 900} = \sqrt{900} = 300 \text{ vars}$$

To find volt amperes with watts and power factor given (800 watts at 0.8 pf):

$$\frac{800}{0.80} = 1000 \text{ volt amperes}$$

To combine two loads of different power factors, first determine if vars are leading or lagging vars. Leading vars subtract from lagging vars. Lagging vars add directly together. Lagging vars are the amount of power that

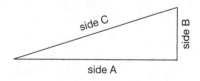

Figure 2-11

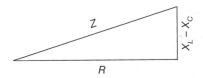

Figure 2-12

is lost due to inductive loads caused by the current lagging the voltage. The more kvars, the less efficient the circuit is. Watts add directly together. To find the volt amperes of the new phase angle, use the above formula.

The impedance triangle has resistance on one side and inductive reactance minus capacitive reactance on the other side. The hypotenuse (the third side) is the impedance. Impedance is the total opposition to the flow of electricity. The equation for this is:

$$Z = \sqrt{R^2 + (XL - XC)^2}$$

In the triangle in Figure 2-12, X_L is inductive reactance and X_C is capacitive reactance.

In Figure 2-13, the first triangle is the impedance triangle. The second triangle is the voltage triangle. Multiply each of a side of the impedance triangle by the amperage I. The third triangle is the power triangle. Multiply the voltage triangle by the amperage again, or I times the impedance triangle. The power factor in the three triangles is the same.

By knowing the amperage and any two sides of any one of the three triangles above, or the pf and one side (either A or C), the following information can be calculated:

Z = impedance

R = resistance

X_L = inductive reactance

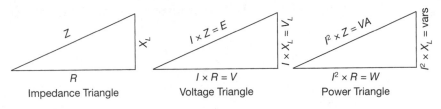

| Impedance Triangle | Voltage Triangle | Power Triangle |

Figure 2-13

X_C = capacitive reactance

pf = power factor

V = voltage

VA = volt amperes

W = watts

vars = reactive volt amperes

Note: If two of the above are given, the rest can be calculated.

Usually, power is expressed in k's, such as kVA, kW or kvars. The k stands for kilo, which equal to 1000 units of whatever is being used. This shortens calculations; for instance,

1,000,000 watts = 1000 kW

500 kVA = 500 × 1000 = 500,000 VA

10 kW = 10 × 1000 = 10,000 Watts

4 kvars = 4 × 1000 = 4000 vars

50,000 watts = 50,000/1000 = 50 kW

Exercise 2-1

1. Convert the following:
   a. 500 watts = how many kilowatts?
   b. 700 kVA = how many volt amperes?
   c. 4000 kW = how many watts?
   d. 1000 kVA = how many volt amperes?
   e. 400 VA = how many kVA?

2. Two power plants are tied together. Power plant A has 1500 kVA at 80% and power factor power plant B has a 700 kW load a 70% pf. Find new the kVA, kW, kvars, and pf.

3. Power plant A has 1000 kVA at 80% pf. Power plant B has 750 kW at 95% pf. Find the new kVA, kW, kvars, and pf when the plants are tied together.

4. a. What is the pf of 1000 kVA, 750 kW?
   b. What is the pf of 750 kVA, 400 kvars?

5. A power plant is at 75% pf 1500 kVA. What leading kvar is needed to bring the pf up to 85%?

Answers to Exercise 2-1

1. A = 0.5 kilowatts
   B = 700,000 volt amperes
   C = 4,000,000 watts
   D = 1,000,000 volt amperes
   E = 0.4 kVA

2. $1500 \times 0.8 = 1200$ kW

   $\sqrt{1500^2 - 1200^2} = \sqrt{810000} = 900$ kvars

   $\dfrac{700}{0.70} = 1000$ kVA

   $\sqrt{1000^2 - 700^2} = \sqrt{510000} = 714$ kvars

   $1200 + 700 = 1900$ kW

   $900 + 714 = 1614$ kvars

   $\sqrt{1900^2 + 1614^2} = \sqrt{6214996} = 2493$ kVA

   $\dfrac{1900}{2493} = 76\%$

   New kVA = 2493

   New kW = 1900

   New kvar = 1614

   New pf = 76%

3. $1000 \times 0.80 = 800$ kW

   $\sqrt{1000^2 - 800^2} = \sqrt{360000} = 600$ kvars

   $\dfrac{750}{0.95} = \sqrt{789^2 - 750^2} = \sqrt{60021} = 245$ kvars

   $750 + 800 = 1550$ kW

   $900 + 714 = 1614$ kvars

   $\sqrt{1550^2 + 845^2} = \sqrt{3116525} = 1765$ kVA

   New kVA = 1765

   New kW = 1550

   New kvar = 845

   New pf $= \dfrac{1550}{1765} = 88\%$

4. $\dfrac{750}{1000} = 75\%$

$\sqrt{750^2 - 400^2} = \sqrt{402500} = 634 \text{ kW}$

$\dfrac{634}{750} = 85\%$

5. Cosine of 0.75 = 41.4 degrees   Tangent of 41.4 degrees =   0.8816
   Cosine of .90 = 25.84 degrees   Tangent of 25.84 degrees = $-$ 0.4843

   $\overline{\hspace{5cm}}$

   0.3973

   $1500 \times 0.75 = 1125 \text{ kW}$ $1125 \times 0.3973 = 447 \text{ kvars}$

2-4   VOLTAGE DROP

Sometimes, a circuit is not adequate for one of several reasons, all resulting in a voltage drop at the load. The following are factors relating to voltage drop:

1. Length of circuit
2. Size of wire
3. Load of the circuit
4. Type of wire (copper or aluminum)

The formula for determining voltage drop is

$$\frac{2KIL}{CM} = VD$$

where
   $K =$ constant: 11 for copper, 18 for aluminum
   $2 = 2$ lengths of wire to the load
   $I =$ amperage
   $L =$ length of circuit one way
$CM =$ Circular mils of wire used

The table for circular mils for wire is in Chapter 9 of the *National Electrical Code®*. The *Code* also allows a maximum of 5% voltage drop from the service to the load, but only a 3% voltage drop between the final overcurrent device and the load (Article 215-2c of the *Code*). On a 120 volt circuit, a 3.6 voltage drop should be maintained.

Example:

100 ft circuit, 110 volts

16 ampere load

#12 THW wire, copper

$$\frac{2 \times 11 \times 16 \times 100}{6530} = \frac{35,200}{6530} = 5.39 \text{ volts}$$

Now try the same problem but this time use #10 copper:

$$\frac{2 \times 11 \times 16 \times 100}{10,380} = 35,200 = 3.39 \text{ volts}$$

The first example using #12 caused too high a voltage drop for the given load. By changing the wire size to #10, the voltage drop was under 3%. Now, cut the load in half—use #12 and 100 ft distance:

$$\frac{2 \times 11 \times 8 \times 100}{6530} = \frac{17,600}{6530} = 2.69 \text{ volts}$$

By cutting the load in half, the voltage drop is also less. The same would hold true if the distance were shortened.

If the voltage is increased, the percent stays the same, but the allowable volts dropped is increased:

Voltage	Allowable 3% drop
120	3.6 volts
240	7.2 volts
480	14.4 volts

Transposing the voltage drop formula adds more dimension and versatility.

Transposing is simply changing the formula around to find out different unknowns. Also, knowing how to transpose eliminates learning six formulas. Remember that in transposing, what is done to one side of the equal sign must be done to the other side.

The basic formula is

$$\frac{2KIL}{CM} = VD$$

Let us find how far we can carry 12 amps on #12 copper at 120 volts while maintaining a 3% voltage drop. First, transpose the formula so that length is on one side of the equal sign by itself:

$$\frac{2KIL}{CM} = VD$$

Then divide each side of the equal sign by $2KI$:

$$\frac{2\cancel{KIL}}{CM} \times \frac{1}{2KI} = \frac{VD}{2KI} \qquad \text{becomes} \qquad \frac{L}{CM} = \frac{VD}{2KI}$$

Now multiply each side of the equal sign by CM:

$$CM \times \frac{L}{\cancel{CM}} = \frac{VD}{2KI} \times CM \qquad \text{becomes} \qquad L = \frac{VD \times CM}{2KI}$$

Now substitute these values $VD = 3.6$, $CM = 6530, 2$, $K = 11$, $I = 12$:

$$L = \frac{3.6 \times 6530}{2 \times 11 \times 12} = \frac{23508}{264} = 89 \text{ feet}$$

You have now transposed the basic formula from finding voltage drop to find the maximum circuit length.

Let us find what wire size is needed on a 16 ampere load, going 100 feet and maintaining a 3% voltage drop on 120 volts. First, transpose the formula, only this time get the CM on one side of the equal sign by itself:

$$\frac{2KIL}{CM} = VD$$

Multiply both sides by CM:

$$\frac{2KIL}{\cancel{CM}} \times \cancel{CM} = VD \times CM \qquad \text{becomes} \qquad 2KIL = VDCM$$

Now divide both sides by VD:

$$\frac{2KIL}{VD} \times 1 = \frac{\cancel{VD} \times CM}{\cancel{VD}} \qquad \text{becomes} \qquad \frac{2KIL}{VD} = CM$$

Now substitute the values $VD = 3.6$, $K = 11$, $I = 16$, and 2 times length of 100 feet:

$$\frac{2 \times 11 \times 16 \times 100}{3.6} = CM \qquad \text{becomes} \qquad \frac{35200}{3.6} = 9777.7 \ CM$$

Next, look in the wire charts at the back of the *National Electrical Code®* (Chapter 9, Table 8). 9777.7 circular mils is more than #12 (6530) but less than #10 (10380). Always use the circular mil in the table that is greater than your calculation. The correct wire size would be #10.

Three-phase voltage-drop calculations use the same formula with a multiplier of 0.866. The multiplier of 0.866 is an unusual number, but all three-phase calculations use the $\sqrt{3}$ or 1.732. If 1.732 is divided by 2, you come up with 0.866. The reason that only half the $\sqrt{3}$ is used is because 2 is already in the formula.

Now let us try a three-phase in a voltage drop problem. A 480 volt, three-phase, 30 ampere load is 380 feet from the source. What size copper wire will be needed to stay at a 3% voltage drop?

First, calculate $480 \times 3\% = 14.4$ volts. Next, transpose the formula to get *CM* on one side of the equal sign by itself:

$$\frac{2KIL \times 0.866}{CM} = VD \qquad \frac{2KIL \times 0.866}{CM} \times CM = VD \times CM$$

$$\text{becomes} \qquad 2KIL \times 0.866 = VD \times CM$$

Divide both sides of the equal sign by *VD*:

$$\frac{2KIL}{VD} = \frac{\cancel{VD} \times CM}{\cancel{VD}}$$

Now substitute the values $VD = 14.4$, $L = 380$, $I = 30$, and $K = 11$:

$$\frac{2 \times 11 \times 30 \times 380 \times 0.866}{14.4} = \frac{217,192}{14.4} = 15,082.8 \ CM$$

15082.8 circular mils is between number 10 and number 8, so number 8 would be the right size (see *National Electrical Code®*, Chapter 9, Table 8).

A quick way to check the voltage drop of a circuit is to use the formula $E = RI$. To use this as a quick check, the wire size must be known. The resistance per 1000 feet is found in the *National Electrical Code®*, Chapter 9, Table 8, Properties of Conductors. The column "ohms/kft" gives the resistance per 1000 feet of the conductor. The columns are divided into coated and uncoated. # 10 coated copper has a resistance of 1.29 ohms per thousand feet. If a circuit draws 16 amperes and is 300 feet long, what is the voltage drop, using # 10 copper coated wire?

$$\frac{2 \times 300}{1000} = 0.600 \times 1.29 = 0.774 \text{ ohms} \times 16 = 12.38 \text{ volts.}$$

2-5   COMBINING LOADS, SINGLE PHASE AND THREE PHASE

In order to calculate a service or feeder, it is sometimes necessary to combine single-phase loads with three-phase loads. It is even sometimes necessary to combine 480 volt equipment with 120/ 208 volt equipment. This looks like adding apples and oranges but really it isn't. Power has one thing in common regardless of voltage or amperage. The watt is what power is all about. 480 volt equipment as well as 120/208 volt equipment can be added together after first being changed to watts. Once added together, it can be divided by the highest voltage and converted back to amperes. This will enable you to size the service or feeder regardless of multiple voltage, single-, or three-phase equipment. Do not, however, try to add amps to amps of different voltages. Also remember that on a transformer changing 480 volt three phase to 120/208 volt, a neutral is present on the low side, but not on the high side. This is because of the delta high side—the neutral is dissipated through the windings. The only neutral at the service is that which was used for the 277 volt lighting.

CHAPTER 2 TEST

1. Is a phase to neutral load single phase?

2. How many watts are there in a 240 volt, three-phase load drawing 10 amperes?

3. In a three-phase, four-wire grounded delta, the high leg is identified by what color?

4. Combine the following loads: (A) Motor load 120 amperes, 480 volt three phase; (B) lighting load of 50,000 watts; and (C) 100 amperes of three-phase, 120 volt load. What is this load in amperes at three phase 480 volts?

5. Power plant A has a 75% power factor using 1200 kVA. Power plant B

has a power factor of 90% using 500 kW. If these power plants are combined, what would be the new power factor?

6. What size copper wire is needed to carry a 120 ampere, 480 volt, three-phase load 1000 feet while maintaining a 3% voltage drop?

7. On a three-phase, four-wire service, the following phase to neutral loads were observed: phase A, 8 amps; phase B, 10 amps; phase C, 3 amps. What is the amperage load on the neutral?

8. If a circuit of 50 amperes is 700 feet long, will # 6 copper coated wire be big enough at 240 volts single phase to maintain a 3% voltage drop?

9. A power plant is running at a pf of 85%. The apparent power is 1200 kVA. What is the kW?

10. What is the kvar in question 9?

11. What is the phase angle in question 9?

12. If a power plant had an output of 1500 kVA at a power factor of 70%, what would the leading kvar be if the power plant were brought up to 85%?

13. What would be the phase angle of the corrected power plant in problem 12?

14. What would be the new kVA in problem 12?

15. What is the amperage of 1600 watts on a 120 volt circuit?

16. What is the voltage of a high leg if the phase-to-phase voltage is 220 volts?

17. A #10 copper single phase is 200 feet long, draws 20 amperes, and has a voltage drop of 13.87 volts. Why?

18. In a panel board, which phase must be designated the high leg?

19. On a high-leg delta system, phase A has 15 amperes between the neutral and phase, and phase C has 5 amperes between the neutral and phase. What is the amperage on the neutral?

20. 13 kV is how many volts?

21. Convert 16,000 watts to amperage at 240 volts three-phase.

22. A 240 volt single-phase motor draws 32 amperes. Convert to watts.

23. A power plant has a 32 degree phase angle. What is the power factor?

24. A power plant is observed to have an output of 900 kW. The apparent power is 1059 kVA. What is the power factor?

25. To improve the power factor, should the phase angle be increased or decreased?

chapter **3**

TRANSFORMERS

3-1 TRANSFORMER BASICS

Transformers are basic electrical components providing a wide range of applications. In some locales, the power company wants to provide only one voltage drop. This may be an impossible situation on some of the intermediate and larger buildings. The cost and size of the wire and conduit in these cases would be prohibitive. Two basic principles to bear in mind concerning transformers are (1) the higher the voltage, the lower the amperage:

$$\frac{100,000}{240\sqrt{3}} = 240$$

but (2) double the voltage, and look at the amperage:

$$\frac{100,000}{480\sqrt{3}} = 120$$

As you can see, we have cut the amperage in half. By reducing the amperage of a circuit, the wire size is reduced since the wire size is based on ampacity and not voltage. Conduit size also may be reduced because of the smaller wire size.

Another advantage is voltage drop. The electrical code recommends a 3% voltage drop from the overcurrent device to the final outlet. A popular voltage is 480/277 four-wire, three-phase service, supplied by the power company. Let us look at an example and compare.

A feeder supplying a lighting panel will handle one hundred 100 watt fluorescent fixtures:

$$100 \times 100 = 10{,}000 \times 1.25 \text{ (continuous load)} = 12{,}5000 \text{ watts}$$

$$\frac{12{,}5000}{120} = 104 \text{ amperes}$$

but

$$\frac{12{,}5000}{277} = 45 \text{ amperes}$$

Now let us look at the voltage drop using 3% in our calculations:

$$3\% \times 120 \text{ volts} = 3.6 \text{ volts}$$

$$3\% \times 277 \text{ volts} = 8.31 \text{ volts}$$

Now, using a number 12 THW, 20 ampere conductor, let us see how far our circuit may go:

$$VD = \frac{2KIL}{Cm} \qquad \text{yielding} \qquad L = \frac{VC \times Cm}{2KI}$$

where $VD = 3.6$ for 120 volts, 8.31 for 277 volts, 2 for load and return, $K = 11$, $I = 16$, and circular mil for #12 THW is 6530 Cm. Substituting our figures:

$$\frac{3.6 \times 6530}{2 \times 11 \times 16} = 66.78 \text{ ft for a 120 volt circuit}$$

$$\frac{8.31 \times 6530}{2 \times 11 \times 16} = 154.15 \text{ for a 277 volt circuit}$$

That is considering the loads to be equal. A standard application for this would be lighting loads, a common light fixture would be a 2 × 4, 200 watt, four-tube fixture. A 277 volt circuit at 16 amperes could handle 22 fixtures up to 154 feet. Conversely, a 120 volt circuit would only to be able to handle 9 fixtures up to 66 feet at the same wattage. That is why it is much more efficient to use 277 volt lighting. The higher-voltage circuit can carry nearly two-and-a-half times more fixtures at more than twice the distance.

However, 480/277 volts cannot satisfy all of our needs. Undoubtedly, we will have a receptacle load and possibly some small areas, such as storage rooms, closets, and so on, for which it would be more practical to use 120 volt incandescent lighting. At this point, we would select a transformer based on whatever load is needed.

The power company will have guidelines that they will follow to select the voltage that will be the most practical as possible. A typical example of this would be:

- Three-phase, 208/120 volt four wire for 10 kVA or more, or at least one 3 horsepower, three-phase motor
- Three-phase, 240/120 volt grounded delta four wire for 50 kVA or more, or at least one 3 horsepower, three-phase motor
- Three-phase, 480 volt three wire for 75 kVA or more
- Three-phase, 480 volt four wire for 100 kVA or more
- Higher voltages may be obtained for higher total connected load

The transformers under discussion are induction-type transformers. Thee transformers have two or more windings tied together by an iron core (see Figure 3-1). The windings of the high-side coil induce a voltage on the low side of the coil windings.

When two or more transformers are tied together electrically and mechanically, certain voltages may be produced. The primary coil is wound around the top of the core and the secondary coil is wound around the bottom of the core. Voltage is induced form the primary coil to the secondary coil. If the primary coil has fewer windings than the secondary coil, then the transformer is a step-up transformer. If the primary coil has more windings than the secondary coil, the transformer is a step-down transformer. A primary coil having twice the windings of the secondary

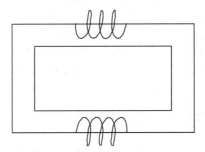

Figure 3-1

winding has a two-to-one turns ratio. The voltage ratio is directly proportional to the turns ratio. This means that the voltage ratio is also two to one. If the primary voltage of the primary coil is 480 volts, the secondary voltage will be half of 480 volts or 240 volts if the turns ratio is two to one:

$$\frac{T_P}{T_S} = \frac{V_P}{V_S} \qquad T_R = V_R$$

Generally, there are several ways of connecting transformers together. We will look mainly at wye-connected and delta-connected transformers and their variations. As in any electrical apparatus, a difference of potential must exist to energize the equipment. This can be done by one phase and one ground, or two phases, or three phases.

Typically, a wye-connected transformer has one side of the coil connected to a phase and other side of the coil connected to ground. This creates a potential difference or voltage on the coil. The potential difference or voltage from this coil is induced in the coil on the secondary side and a potential difference is created, depending on how the lower coil is connected on the secondary side of the transformer. We will look at popular combinations of transformers later.

Delta-connected transformers derive their potential difference phase to phase. There are variations to this, as we will discuss later. Transformer connections on the high side are independent of connections on the low side. In other words, you may use a delta connection on the high-side winding and a wye connection on the low-side winding. In this discussion, we will consider six of the more popular combinations and concentrate on these. These combinations are:

Primary Windings	Secondary Windings
Wye	Wye
Wye	Delta
Delta	Wye
Delta	Delta
Delta	High-leg delta
Wye	High-leg delta

3-2   DELTA AND WYE CHARACTERISTICS

Before we examine each of these connections, let us look at some electrical characteristics of delta and wye connections. Since, by earlier discus-

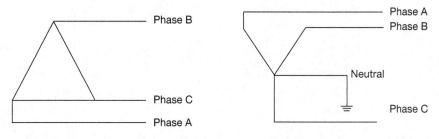

Figure 3-2

sion, we know that wye connections are phase to ground, it follows that phase-to-ground voltage times $\sqrt{3}$ equals phase-to-phase voltage:

$$E_L \times \sqrt{3} = E_P$$

Amperage stays the same: $I_L = I_P$. Delta connections are phase to phase, so it follows that $E_L = E_P$. However, amperage is $I_L \times \sqrt{3} = I_P$.

Relationships between transformers are shown in Figure 3-2.

3-3   DELTA CONNECTION

The delta connection is broken down into transformer windings. Each transformer winding is represented by one side of the delta triangle, and each angle is formed by two transformer windings joined together. By numbering each end of the windings, 1–6 (two ends to each winding), we will transpose the windings from the triangle into a side-by-side relationship (see Figure 3-3).

To tie the windings together, jumper conductors are placed at points 2 and 3 (see Figure 3-3), at points 4 and 5, and finally at points 1 and 6, resulting in the arrangement shown in Figure 3-4. Now the jumpers for the

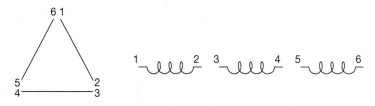

Figure 3-3

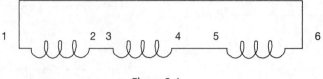

Figure 3-4

phases can be added, as shown in Figure 3-5. The delta connection is now complete.

3-4   WYE CONNECTION

The wye connection is broken down into transformer windings (see Figure 3-6. Each winding is represented by a line. The three windings join together at the apex of the wye. A common tie is formed at this point, as in the Delta, and the windings are numbered 1–6 and placed in a side-by-side relationship (see Figure 3-7).

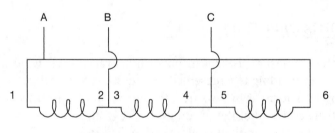

Figure 3-5

Figure 3-6

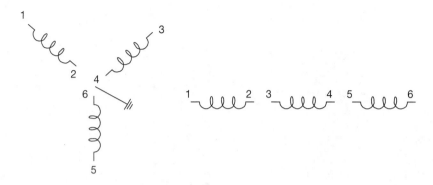

Figure 3-7

Now all even numbers are joined together, 2, 4, & 6, resulting in the arrangement shown in Figure 3-8. The jumpers to the phases can now be added (see Figure 3-9). The wye connection is now complete.

In delta connections, the voltage is the same across the winding and between each winding. For example, 480 volts are supplied to the high side of the delta-connected windings. The voltage between phases is 480; the voltage across the winding is 480.

The voltage on a wye connection is different. The voltage between phases is still 480, but the voltage across the winding is divided by $\sqrt{3} =$

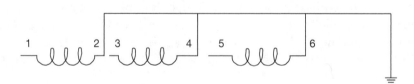

Figure 3-8

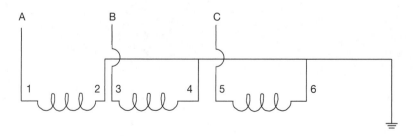

Figure 3-9

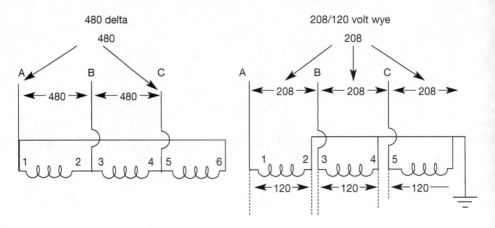

Figure 3-10

277 volts. Remember, the voltage across the winding is the voltage phase to ground. Another example of the wye-connected and delta-connected relationship is 480 delta and 208/120 volt wye shown in Figure 3-10.

3-5 GROUNDED DELTA

The grounded delta (Figure 3-11) is another popular connection. This provides three-phase power and two phases to ground for 120 volt lighting. The only difference between the straight delta and the grounded delta is the grounding of the center winding, usually of the center transformer. The voltages are as shown in Figure 3-12. The third phase to ground is

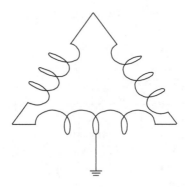

Figure 3-11

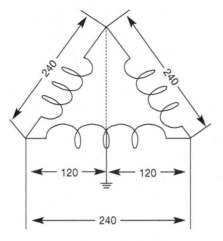

Figure 3-12

called the high leg. It has a higher voltage to ground than the other two, which can be shown two ways (see Figure 3-13).

First, take the delta triangle and add a line from the top of the triangle to the ground. Now we have two right triangles. Each side of the new triangle is measured in volts. The bottom side of the triangle is 120 volts long. The side opposite the bottom side is 240 volts long (the hypotenuse). Now, using the Pythagoran theorem for a right triangle we have

$$A^2 + B^2 = C^2 \qquad \text{or} \qquad 240^2 - 120^2 = B^2 \qquad \text{or}$$

$$57600 - 14400 = \sqrt{43200} = 207.6 \text{ volts}$$

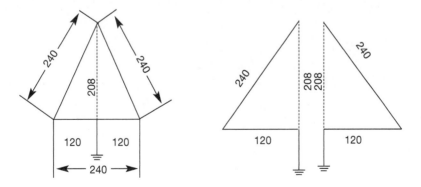

Figure 3-13

Another way to arrive at the voltage of the high leg to ground is to multiply the 120 volts by $\sqrt{3}$ or 1.732, resulting in 207.6 volts. Putting the above triangle together will give three-phase, 240 volt, two 120 volt lighting legs, one 208 volt high leg to ground, and one neutral.

A high-leg delta can be connected as two or three transformers. The two-transformer hook up has a reduced rating of 57.7% of the three transformer hookup. The two-transformer hookup is called open delta. The three-transformer hookup is called closed delta. For example, three 100 kVA transformers have a combined rating of 300 kVA.

3-6   OPEN DELTA

An open delta using two 100 kVA transformers would have a rating of 86.6% of the sum of the kVA of the transformers being used (see Figure 3-14. Or, if two identical kVA rated transformers are added to a third, or phantom, transformer, giving a combined 300 kVA, the rating of the transformer bank would be 57.7% of the 300 kVA:

$$200 \text{ kVA} \times 86.6\% = 173 \text{ kVA}$$

$$300 \text{ kVA} \times 57.7\% = 173 \text{ kVA}$$

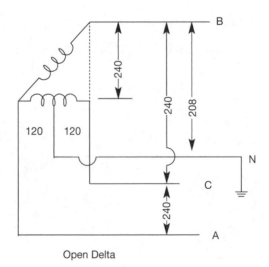

Open Delta

Figure 3-14

An open delta transformer bank is used on limited three-phase loads, where most of the load is going to be 120/240 volt single phase. A good example of this would be a service station that has a three-phase air compressor and the rest of the load would be 120/240 volt single phase. The power company benefits by having to supply two transformers instead of three.

3-7   MOTOR CONTROL TRANSFORMER

This circuit shown in Figure 3-15 is good for motor control or limited lighting. The secondary side can be changed to delta, so let us look at the secondary winding in detail. In Figure 3-15, the windings have been stacked or paralleled.

The circuit in Figure 3-16 would be used for heavy motor loads on the 480 volt three-phase. Lighting is obtained from the secondary 120/208 volt wye. 120 volt load can be balanced between all phases to ground.

3-8   SERIES AND PARALLEL WINDINGS

The secondary side can be changed to delta so let us look at the secondary winding in detail. In the example in Figure 3-17, the windings have been

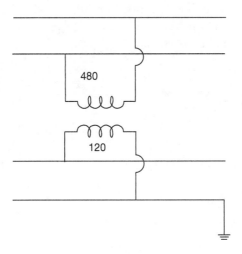

Figure 3-15

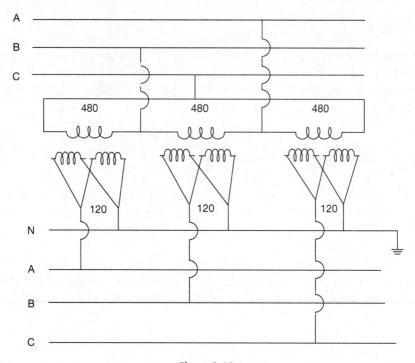

Figure 3-16

stacked or paralleled. X_1 and X_3 are tied together and X_2 and X_4 are tied together and grounded. Put together with other windings connected in the same manner, the X_1–X_3 connection of the first set of windings has a phase-to-phase relationship of $\sqrt{3}$ times the phase-to-ground voltage of 120 volts, resulting in 208 volts.

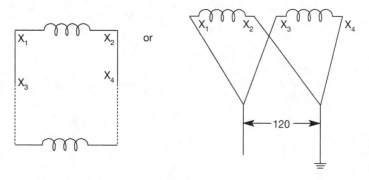

Figure 3-17

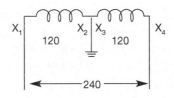

Figure 3-18

The same windings can be connected in series, as shown in Figure 3-18. Here X_2 and X_3 are joined together and grounded. Joined with other windings connected in a like manner, an open or closed delta is formed. The disadvantage of this is that only two 120 volt legs are available, versus three 120 volt legs available on the three-phase, four-wire wye system.

Figure 3-19 shows a three-phase, four-wire grounded delta with a 480 volt Primary. This connection has two legs for lighting and three legs for three-phase power.

The transformer nameplate must be read before tying transformers together in any connection. Nameplate data will tell kVA, primary and secondary voltage, and type of connection—wye or delta for both the primary winding and secondary windings. Impedance is another factor that is on the nameplate. This is expressed in percentage and represents the core loss and copper loss through the transformer windings. When tying two or

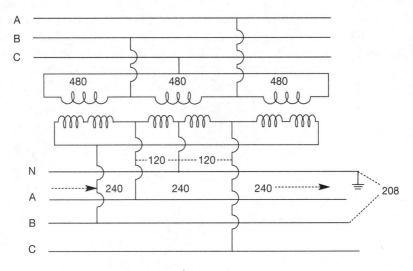

Figure 3-19

three transformers together, the impedance value should always be plus or minus 0.5% of each other. This will assure that circulating currents will not overburden one winding over another, causing a premature failure.

Another important check is to make certain in two- or three-transformer connections that the windings are the same in each transformer. If one transformer winding is in series, all must be in series.

Tap changers are available on transformers which merely add or subtract voltage on the primary side. There is usually a center position that is 100%; for instance, 480 volts. There are five positions at 2.5% increments off center, which results in 97.5% and 95%; 102.5 % and 105%. This is for changing the voltage if the input voltage is too high or too low. All taps in other than single-transformer installations must be set to the same percentage.

3-9   TRANSFORMER POLARITY

Polarity is another important consideration when putting two or three transformers together. Usually, this is indicated on the nameplate; however, transformers are sometimes repainted and the polarity markings will not show. Transformers of opposite polarity will not go together. Transformers have either additive polarity or subtractive polarity. Usually, the H_1 bushing is located on the left-hand side facing the front of the transformer. The X_1 bushing is dependent on whether the transformer is additive or subtractive. If the X_1 bushing is on the same side as the H_1 bushing, the transformer is additive. To check for polarity; first tie the H_1 bushing to the adjacent X bushing. Connect a 6 volt source between the X bushings (see Figure 3-20). Next read with a meter between the other X bush-

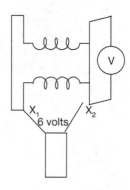

Figure 3-20

ing and H bushing. If the secondary voltage subtracts from the primary, the transformer is subtractive and the X_1 bushing is on the opposite side of the H_1 bushing. Note: the source voltage should be lower than normal full voltage. If the voltmeter reads 24 volts, the transformer is additive. If the voltmeter reads 12 volts, the transformer is subtractive. To change the polarity of a transformer, reverse the leads on either the primary or secondary sides, but not both sides.

3-10   BUCK AND BOOST TRANSFORMERS

Transformers are an integral part of a large electrical system. They can be used for doorbells, motor control circuits, and separately derived systems to name a few applications. Occasionally, voltage is a little too high or too low. A piece of equipment will burn up at 208 volts if it is designed for 240 volts. Note most equipment will run well on lower voltage, but will just use more amperage. However, some equipment will need to be at the rated voltage. To correct this problem, a boost transformer is needed. A transformer is added for either a 5% increase or a 10% increase for a boost of 12 volts or 24 volts, respectively.

When a 5% increase is needed, the secondary windings are paralleled or stacked (see Figure 3-21). When a 10% increase is needed, the secondary windings are put in series. If the voltage is too high, then the buck

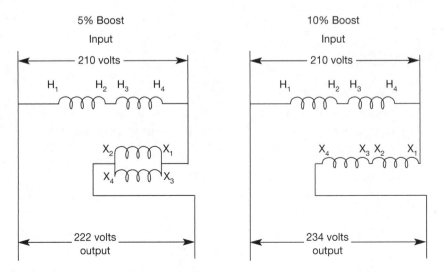

Figure 3-21

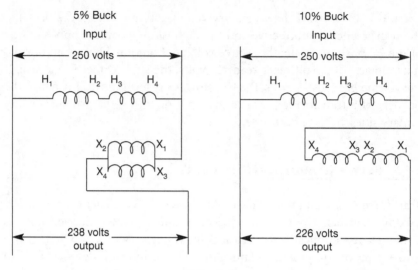

Figure 3-22

connection is used (see Figure 3-22). By reversing the polarity of the secondary connections, the voltage is subtracted, instead of added as in the previous example.

These connections allow more flexibility in the electrical installations. Again, when a 5% decrease is needed, the secondary winding is paralleled or stacked. When a 10% decrease is needed, the secondary winding is put in series.

CHAPTER 3 TEST

1. What is the purpose of a transformer?
2. A phase-to-neutral reading of 265 volts was observed on a wye-connected transformer bank. What is the phase-to-phase voltage?
3. What kind of transformer connections are in series?
4. What kind of connections are in parallel?
5. An open-delta transformer has two 15 kVA transformers tied together. What is the output of the bank?
6. What is the amperage of a transformer bank that has three 25 kVA transformers wye-connected at 208/120 volts?
7. What is the output of three delta-connected 25 kVZ transformers?

Figure 3-23

8. Show the connections of a wye-connection for the transformers in Figure 3-23.

9. A 10 kVA transformer has how many amperes available at 240 volts?

10. If the same transformer were wye-connected, how many amperes would be available?

11. What does a tap changer do?

12. If the X_4 bushing is on the same side as the H_1 bushing the transformer would be ___?

13. Does 5% buck transformer have the secondary coil connected in series or parallel?

14. Does 10% boost transformer have the secondary coil connected in series or parallel?

15. What is transformer impedence?

16. Connect the transformer bank in Figure 3-24 primary 120 volt/208 wye to secondary 480 volt delta as a step-up transformer.

17. If a transformer has a 30-to-1 turns ratio and the secondary is 240 volts, what is the primary voltage?

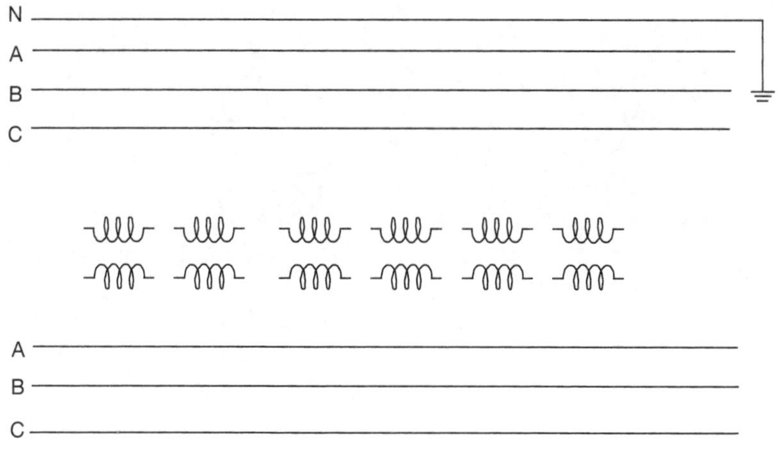

Figure 3-24

18. In a wye-connected transformer bank, how many coils are grounded?

19. In a delta-connected transformer bank, how many coils are connected?

20. How many transformers are required for an open-delta transformer bank?

chapter 4

MISCELLANEOUS CALCULATIONS

4-1   CONDUIT SIZE—TWO OR MORE CONDUCTORS OF THE SAME SIZE

There are 58 pages of tables in the *National Electrical Code®* (NEC) covering applications in which conductors are the same size and have similar insulation (Annex C).* On the first page, the tables are listed by raceway, starting with Table C1, Electrical Metallic Tubing (EMT). There are twenty-four tables in all. Every other table is for compact conductors and are designated by an "A." The conduit fill should not exceed 40% of the cross-sectional area of the conduit. Conduit fill is based on Table 1 Chapter 9 of the NEC. If compact conductors are used in EMT conduit, the appropriate table would be C1A. If rigid conduit is used for compact conductors, the appropriate table would be C8A.

There is a table for each of the above. Find the correct table for the raceway being used, then find the type and size of wire being used. For instance, four 1/0 copper THW conductors are installed in a rigid metal raceway. What size conduit would be needed? The procedure is

1. Look up Chapter 9 Table C8 on page 693.
2. Find the type of wire in the far-left column.

*This chapter contains material reprinted with permission from NFPA 70-2005, *National Electrical Code®*, Copyright © 2004, National Fire Protection Association, Quincy, MA 02269. This reprinted material is not the complete and official position of the NFPA on the referenced subject, which is represented only by the standard in its entirety. National Electrical Code and NEC are registered trademarks of the National Fire Protection Association, Quincy, MA.

3. Find the wire size in that group in the next column.

4. Go across the column until the number of conductors is equal to or greater than the number of conductors being installed.

5. When the number has been found (in this case, six, which is more than the four to be installed), look to the top of the column and find the conduit size, in this case 2 inches.

Another example would be nine #12 THHN conductors in a ½" EMT conduit. Look up the in Chapter 9 Table C1 under THHN conductors and find the number of #12s allowed in a ½" EMT conduit. The number allowed is nine. The circuit makeup of these wires are two sets of three-phase conductors with a neutral for each set, commonly referred to as two full boats. The ninth wire is a ground wire and is not counted as a current carrying conductor.

The purpose or end use of the circuit affects the ampacity of the circuit. If the circuit is used for commercial lighting (fluorescent) or computer-type circuits, the neutral becomes a current-carrying conductor. This would make eight of the nine conductors current carrying and the circuit would need to be derated according to Table 310.15(B)(2)(a) to 70%. This would mean that the #12 conductors could only carry 14 amperes. The reason for the reduced amperage is to reduce heat generated by the higher ampacity. Many larger commercial jobs require that home runs be #10 instead of #12. #10 derated to 70% would be 21 amperes. It also allows less voltage drop.

If the above circuit makeup were to serve general-use receptacle loads, the two neutrals would be used only to carry the unbalanced load and would not count as current-carrying conductors. Now there are only six current-carrying conductors and the circuit would only have to be derated to 80% or 16 amps. Note that all nine conductors must be counted in the conduit fill.

4-2 DIFFERENT-SIZE CONDUCTORS

In a less than perfect world, the conductors are not the same size in a raceway and different set of tables is needed to determine the raceway size. The first step would be to go to Chapter 9 Table 5, Dimensions of Insulated Conductors and Fixture Wires, to find the correct wire size and insulation. Table 5 gives the dimension of the conductor with its insulation. The other table that will be used is Chapter 9, Table 4. Chapter

9, Table 4 is arranged by article number—the type of raceway being used. For example, rigid metal conduit is found under Chapter 9, Table 4, Article 344.

This table gives the maximum square inches of conductor that can fill the conduit. There is a separate table for each type of conduit—EMT and rigid metal conduit. It is important that the right table be used for conduit fill. The table is further broken down by the number of wires in a conduit. The one most often used is two or more conductors with 40% fill. The other columns can be used as the need arises. The last column is 60% fill. This column can be used when determining conductor fill for nipples not more than 24″ long. Also note that conductors in a short nipple with 60% fill do not have to be derated because of the number of conductors.

Problem

What size EMT conduit would be needed for three #6 THHN conductors and one #8 THHN conductor? First, find in THHN conductors Chapter 9, Table 5, then look for #6. Follow across to the in² column and find 0.0507 square inches. Now, look up #8 THHN in the same table. The answer is 0.00366 square inches. Next, multiply each size conductor by its square inches and add them together:

$$\#6 = 0.0507 \times 3 = 0.1521$$
$$\#8 = 0.0366 \times 1 = 0.0366$$
$$\overline{}$$
$$0.1887 \text{ square inches}$$

Now, go to Chapter 9, Table 4 and find the column for over two wires and 40%. Look until a number bigger than 0.1929 is found in the square inch column. That would be 0.213 square inches. On the same line, look under trade size and find the conduit size, in this case, ¾″.

In another example, there are three THW 4/0, one 1/0 THW, and one #4 THW conductors in an EMT conduit. Go to Table 5 and look up THW conductors:

$$4/0 = 0.3718 \times 3 = 1.1154$$
$$1/0 = 0.2223 \times 1 = 0.2223$$
$$\#4 = 0.0973 \times 1 = 0.0973$$
$$\overline{}$$
$$1.436 \text{ square inches}$$

Now look in Chapter 9, Table 4, EMT in the column for over two wires and 40%. Chapter 9 Table 4 is arranged by the article number of the type of raceway being used. For instance, the right data for EMT would be found in Chapter 9, Table 4, Article 358. Find the number larger than 1.436 square inches. The answer is 2.343 square inches. Look under trade size for the conduit size and the correct size is found to be 2½" EMT conduit.

Often, parallel conductors are split between two or more conduits. As long as the makeup of the circuit follows the rules of 310-4, this is permissible. There are advantages to keeping the size of the conductor down. For instance, consider a three-phase four-wire circuit paralleled with six 4/0 THW conductors paralleled for the ungrounded conductors, two 1/0 THW conductors paralleled for the neutral, and one #4 THW ground wire. What size EMT raceway will be needed for this circuit?

$$6 \times 0.3718 = 2.208$$
$$2 \times 0.2223 = 0.4446$$
$$1 \times 0.093 = 0.093$$
$$\overline{}$$
$$2.7684 \text{ square inches}$$

This circuit would require a 3" EMT conduit. If this circuit used the neutrals as current-carrying conductors, that would cause a derating of 70% (eight current-carrying conductors or 70% of 460 amperes = 322 amperes). If the circuit is split between two conduits, the conduit size would change as well:

$$3 \times 0.3718 = 1.1154$$
$$1 \times 0.2223 = 0.2223$$
$$1 \times 0.093 = 0.093$$
$$1 \times 0.093 = 0.0930$$
$$\overline{}$$
$$1.4307 \text{ square inches}$$

The closest size EMT conduit is 2½". Now the circuit has four current-carrying conductors in two conduits and only needs to be derated to 80% or $460 \times 0.80 = 368$ amperes.

Note: Since electricity travels on the surface of a conductor rather than through it, increasing the size of the conductor only serves to give more surface. Stranded conductors have more surface than solid conductors. Doubling the circular mil area of a conductor does not double the amperage.

For example, consider 500 MCM THW conductors compared to two conductors in parallel, each with 250 MCM THW conductors. 500 MCM THW conductors have an ampacity of 380 amperes. 250 MCM THW conductors have an ampacity of 255 amperes. When put in parallel, the 250 MCM conductors have an ampacity of 510 amperes compared to the 380 amperes of the 500 MCM THW conductors. This is because even though the circular mil is the same, there is more surface area around the two 250 MCM conductors. In Chapter 9, Table 8 of the NEC Conductor Properties, 250 MCM has a diameter of 0.572 inches, whereas 500 MCM has a diameter of 0.813 inches. The total area of 250 MCM is 0.260 square inches and the total area of 500 MCM is 0.519 square inches, or twice as much. However, if the surface or circumference of the conductor is used 2 × R × 3.14 = 1.80 inches for 250 MCM and 2 × 0.4065 × 3.14 = 2.55 square inches. two 250 MCM's would have a surface area of 2 × 1.80 inches or 3.60 inches compared with 500 MCM of 2.55 square inches.

4-3   SIZING JUNCTION BOXES FOR #6 AND SMALLER CONDUCTORS

Sizing of outlet boxes, pull boxes, junction boxes, and conduit bodies is covered in Article 314. For #6 and smaller conductors, there are two ways to calculate conductor fill. The first is with all of the conductors the same size. There are two tables for sizing boxes, 314.16A and 314.16B. 314.16A gives standard box sizes, the volume in cubic inches, and the number of conductors allowed by size. 314.16B is the volume allowance for each size conductor.

Every conductor that comes into the box and is spliced in the box counts as one conductor. If the conductor passes through the box, it will be counted as one conductor. If a conductor is contained within the box, it does not count.

Devices inside the box affect the volume allowances and conductor count. Internal clamps, fixture studs, or hickeys will cost one conductor or one volume allowance of the largest conductor for each device. A double volume allowance is required for each yoke or strap with one or more devices. A single volume allowance is required for the largest ground wire.

There are several combinations of boxes and rings that can be used. For example, a ¾″ EMT conduit enters a four-square box 1½″ deep. The total number of #12 conductors allowed in this box according to Table 314.16A is nine. This is the maximum number allowed without volume allowances. This would accommodate two full boats (three phases, one neutral) and a ground with no splices. If a ½″ flex is added for a fixture whip, the count would change to two in for the whip, two conductors in from the home run, two conductors out, and five conductors straight through. The ground splices, but only counts as one conductor. Now there are 12 conductors and the box is too small without an extension. A larger box—4 × 4 × 2⅛″ will accommodate 13 #12's and will work well.

If this box is in the wall is for an outlet box and one receptacle is going to be installed, two conductors would be spliced from the home run and exit to continue on. Two volume allowances would have to be subtracted for the receptacle, which means two less conductors would be allowed. If a ½″ raised device ring is used (it has a 3 cubic inch allowance), it can be added to the volume of the box. If the receptacle is pigtailed and does not leave the box, it does not need to be counted in the conductor count.

Different size conductors are calculated by volume allowance in Table 314.16B. Three #10 and three #12 in a conduit need a box. The volume allowance for #10 is 2.50 cubic inches and for #12 it is 2.25 cubic inches. Multiply the number of conductors by their respective volume allowances and add them together:

$$3 \times 2.5 = 7.5$$
$$3 \times 2.25 = 6.75$$
$$\overline{}$$
$$14.25 \text{ cubic inches}$$

A box with a minimum volume of 14.25 cubic inches is needed. If this is a pull box or junction box with no other allowances required, a four-square by 1¼″ box will work out fine, but if a tap to a light is needed for the #12s, the count would change. Add a ½″ flex to the lighting fixture, and two conductors are spliced, two conductors come in from the home run, and two go out to continue. One #12 and three #10s go straight through:

Flex	two #12 × 2.25 = 4.5
In	two #12 × 2.25 = 4.5
Out	two #12 × 2.25 = 4.5
Straight through	1 #12 × 2.25 = 2.25
Straight through	3 #10 × 2.5 = 7.50
	23.25 cubic inches

Now a box is needed that has at least 23.25 cubic inch volume. If a four-square by 1½″ box were used, an extension would be needed. The original box would be good for 21 cubic inches (Table 314.16A):

Needed: 23.25 cubic inches
Original: 21 cubic inches

2.25 cubic inches

An extension ring of at least 2.25 cubic inches would be needed.

Conduit bodies such as T's, Lbs, and the like must have a minimum of twice the cross-sectional area of the conduit they are attached to.

4-4 JUNCTION BOXES FOR #4 AND LARGER CONDUCTORS

A junction box in a straight pull of #4 and larger conductors must be eight times the largest conduit to the opposite wall. If a junction box contains one 3″ conduit, the length of the junction box must be 3 × 8 or 24″ to the opposite side. The width of the box only has to accommodate the lock nut and bushing. If more conduits are entering the same face as the 3″, the width would have to be wide enough to accommodate the additional conduits and include room for locknuts and bushings. The length would be 3″ × 8 or 24″.

An angle pull for conduits with #4 or larger conductors has to be six times the trade size of the largest conduit plus the sum of the diameters of the remaining conduits. The spacing coming into the box should allow enough room for locknuts and bushings. However, if spacing has not been determined it should be considered between the conduits so that a small adjustment will not have to be made when the field installation is made.

An angle pull for conduits with #4 or larger conductors has to be six times the trade size of the largest conduit plus the sum of the diameters of the remaining conduits. This is true for the wall of the box that is entered as well as the opposite wall of the box that is being exited. This means that the length and width of the box must be calculated. In addition, the distance between the conduits must be a minimum of six times the conduit size. For example, a junction box is needed for one 3″ and two 2″ conduits. What size junction box would be needed?

$$3 \times 6 = 18''$$
$$2 + 2 = 4$$

$$22''$$

The length and width of this junction box would be a minimum of 22″ × 22″.

4-5 OVER 600 VOLTS

To size junction boxes for circuits over 600 volts, instead of sizing the junction box by the largest conduit, the size is determined by the outside diameter of the oversheath of the largest cable or conductor. For a straight pull, the outside diameter oversheath of the largest cable or conductor is multiplied by 48 times. If none of the cables or conductors are shielded or lead covered, the distance can be shortened to 32 times the outside diameter of the largest nonshielded or cable.

For angle or U pulls with conductors over 600 volts, the distance is 36 times the outside diameter of the oversheath of the largest cable or conductor. This distance should be increased by the outside diameter oversheath of additional conductors or cables when entering the same wall of the box. For nonshielded and nonlead-covered cables and conductors, this distance can be reduced to 24 times the outside diameter of the largest cable or conductor plus the outside diameter of additional cables or conductors. Even though conduits enter the box, the conduit itself is ignored under this set of rules.

4-6 SHIELDED CABLE

A shielded cable can be broken down in six basic components (see Figure 4-1):

A) Conductor

B) Semiconductor

C) Insulation

D) Semiconductor

E) Shielding

F) Outer protective jacket

The conductor (A)is either compact or regular. The semiconductor (B) is extruded on to fill in between the strands of the conductor and the insulation. The conductor side of the semiconductor is ribbed and fits snuggly between the outer strands of the conductor; the insulation side is nearly smooth, although textured. This layer of semiconductor fitting between the strands on the conductor side and nearly smooth on the insulation side has the effect of making the electrons "think" that the conductor is

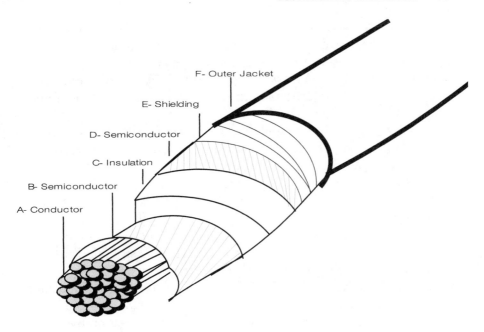

F- Outer Jacket

E- Shielding

D- Semiconductor

C- Insulation

B- Semiconductor

A- Conductor

Figure 4-1　Shielded cable construciton.

solid and smooth, not stranded. This stops stress from building up between the conductor (C) and the insulation.

The insulation layer is next. It is the thickest part of the cable assembly. There are a variety of insulations used. One example is cross-linked polyethylene; another is ethylene–propylene (EPR). Cross-linked polyethylene is very inflexible and difficult to cut with a knife. Ethylene–propylene is a rubber-based synthetic of the polyethylene family. This insulation is considerably more flexible than cross-linked polyethylene, as well as much easier to work with. Its elasticity allows it to contract and expand when high voltage is applied. When this insulation is sanded, it is soft and pliable, like a big eraser.

The next layer of the cable assembly is another semiconductor (D). This is the transition between the insulation and shielding. The purpose of this second semiconductor is to ensure that a tight bond between the insulation and shield is maintained. Without it, going around a 90° bend would result in little separation between the shielding and the insulation. If the semiconductor weren't there, an air pocket or void would be formed. This void would cause the electrons to accumulate in the void, which would cause an electric discharge (corona discharge). This corona discharge along the cable jacket or between the cable jacket and the race-way or other materials would ionize the air, forming ozone gas. This gas is just as conductive as copper and will cause a chemical change in the jacket. Once arcing starts, it will continue until the cable is destroyed. The semiconductor is elastic and flexible enough to prevent a void from occurring.

The next layer in the assembly is the shielding (E). The shielding drains off the excess electrons to a safe level. It controls the radial stress between the insulation and the copper shield. With the insulation be-tween the conductor and the shielding, a capacitor is formed. The flux lines that leave the conductor will leave at right angles to the conductor (this is called radial stress) if the conductor is perfectly round. Since the first semiconductor (B) gives the stranded conductor the appearance of a solid conductor, this causes equal and symmetrical distribution around the conductor, putting an equal amount of stress on the insulation. If the stranded conductor didn't have the semiconductor on the conductor, the flux lines would leave at right angles to each strand and would not be even or symmetrical. This would result in uneven stress on the insu-lation. Now that even stress has been achieved, it can be contained by the insulation.

Another kind of stress is tangential stress. This stress results as the voltage is of a circuit rises from zero to its peak value and then drops to its maximum peak in the opposite direction ($13,000 \times 1.414 = 18,382$ volts up and 18,382 volts down, 60 times a minute). The equipotenial lines vary in the insulation and can cause a whipping effect on the elec-trons on the surface of the cable. This whipping effect could cause the surface electrons to arc across the raceway that the cable was installed in. When semiconductors between the insulation are used, the equipotential lines level out and the voltage is evenly distributed throughout the insu-lation.

The third type of stress is longitudinal stress. Since a cable isn't sus-pended in the center of the raceway, or lying perfectly flat on the bottom, in some places it is touching the raceway and in other places it is not touching. The cable could be dirty or the conduit could contain water. If an unshielded cable were used, the accumulation of electrons on the jack-et would look for a place to discharge, either between two places on the jacket or between the jacket and the raceway. A number of conditions can exist that would cause a static charge to discharge. A shielded cable elimi-nates many of these problems. The shielded cable holds the stress within the insulation. Since the shield is grounded, the electrons that collect are drained off through the ground harmlessly. Although the electrons are not completely eliminated, they are drained to a safe level.

After the stress cone is put onto the end of the cable, it usually terminates to an unshielded cable or busbar. In either case, it is isolated from surrounding metal parts by barriers, usually bakelite or similar material, so that the cable or busbar is not stressed by surrounding metal parts.

The final part of the cable assembly is the jacket (F), used to physically protect the cable. The jacket is similar to 600 volt insulation.

The stress cone or cable termination does not relieve the stress at the end of the cable. Instead, it dissipates the concentration of flux lines, equipotential lines, and longitudinal stresses caused by the abrupt ending of the shield. It also provides a place to make the ground connection to keep the shield at ground potential.

4-7   GROUNDING ELECTRODE CONDUCTOR

Acceptable grounding electrodes are:

- Underground metal water pipe at least 10′ in direct contact with the earth.
- The metal frame of a building is acceptable if it is effectively grounded.
- Concrete-encased electrode. Must be encased with at least 2″ of concrete. Rods must be at least ½″ in diameter and at least 20′ in length, and be made of electrically conductive material. Copper conductor must be at least 20′ long and at least #4 AWG.
- Ground ring. Must be at least 20′ long in direct contact with the earth and a minimum of #2 copper.
- Rod and pipe electrode, commonly referred to as ground rods. Must be at least 8′ in length. The diameter of the rod depends on the type of rod being used. ¾″ is acceptable for pipe and conduit electrodes, the outer surface coated with corrosion-resistant protection.
- Plate electrodes. The plate must be at least 2 ft² and in direct contact with the earth. Iron and steel plates must be at least ¼″ thick. Plates of nonferrous metal must be at least 0.06 inches or just under 1/16″ thick.

Conductor sizes for grounding electrodes are as follows:

1. Underground metal water pipe (Table 250.66)
2. Metal frame of building or structure (Table 250.66)

3. Concrete encased electrode, maximum #4 copper AWG

4. Grounding not smaller than #2 copper AWG

5. Rod and pipe electrodes, maximum #6 copper AWG

6. Plate electrodes, maximum #6 copper AWG

Table 250.66 sizes the grounding electrode conductor by the size of the largest ungrounded conductor. This table can also be used for separately derived systems (transformers).

If the resistance of a rod, pipe, or plate electrode doesn't have a resistance of 25 ohms or less to ground, one more electrode must be added, but they must be at least 6′ apart.

4-8   EQUIPMENT GROUNDING CONDUCTOR

The equipment grounding conductor must have an outer covering of green or green with one or more yellow stripes. For conductors larger then #6 AWG, copper can be identified permanently at each end and every point that is accessible by stripping the insulation or covering of all of the exposed conductor, or coloring the exposed part of the insulation green, or by using green tape or green adhesive labels.

There are several types of equipment grounding conductors, including metal raceways that are acceptable:

- Conductors of nearly every type, whether insulated, covered, or bare. A busbar of any shape.
- Metal conduit, including rigid metal conduit, intermediate metal conduit, or electrical metallic conduit.
- Flexible metal conduit that is listed and used with listed fittings for grounding according to Article 250.118(5) a–d.
- Liquid-tight flexible metal conduit that is listed according to Article 250.118(6) a–e.
- Flexible metallic tubing that uses listed grounding fittings according to Article 250.118 (7) a–b
- Armor of AC cable according to Article 320.108.
- The copper sheath of mineral-sheathed cable.
- Type-MC cable that is listed and identified for grounding according to Article 250. 118 (10) a–b.
- Cable trays according to Article 392.3 and 392.7.

- Cable bus framework according to Article 370.3
- Other electrically continuous metal raceways and auxiliary gutters listed for grounding.

Most larger commercial and industrial jobs require using a copper conductor or an equipment-grounding conductor instead of a raceway. To size an equipment-grounding conductor, use Table 250.122. This simply sizes the equipment-grounding conductor by the overcurrent device instead of the ungrounded circuit conductors, but it cannot be larger than the ungrounded circuit conductors.

Oftentimes, more than one circuit is in a raceway. If the circuits are of different sizes, only one equipment-grounding conductor is required and that is sized for the largest overcurrent device involved in the raceway (Table 250.122).

If the ungrounded conductors increase in size, the equipment-grounding conductor must be increased in size according to the circular mil area of the ungrounded conductors.

The equipment-grounding conductors for parallel conductors in multiple raceways must have an equipment-grounding conductor in each raceway sized according to Table 250.122.

When the overcurrent device consists of an instantaneous-trip circuit breaker or a motor-short circuit protector (Table 430.52), the equipment grounding conductor size can be based on the rating of the motor-overload protective device (Table 250.122).

Since conductors for motors are generally sized at 125% of the full load current of the motor and branch circuit, short circuit, and ground fault protection ranges from 150% to 800% for an instantaneous trip breaker, it would be very possible to size the equipment-grounding conductor larger than the ungrounded conductors. It is appropriate to size the equipment-grounding conductor by the motor overload protection, which runs between 115% and 125% (Table 430.32A1). This is more in line with the wire size than the overcurrent device.

A 30 horsepower induction-type, squirrel-cage motor at 220 volts has a full load current of 80 amperes. If an instantaneous breaker were used (Table 430.52), 800% of 80 is 640 amperes and a 700 ampere breaker would be required. If the breaker were not sufficient for the starting current of the motor, the instantaneous-trip circuit breaker could be increased up to 1300%, or 13 × 80 = 1040, which would require a 1000 ampere breaker and an equipment-grounding conductor of 2/0 copper, much larger than the #3 copper ungrounded conductors used to supply the motor. Since the size of the overload protection can be used, 125% × 80 = 100 amperes, and a #8 copper conductor can be used.

4-9   NEUTRAL SIZE

Sizing the neutral or grounded conductor is done according to Table 220.61. The neutral acts as a return conductor for phase-to-neutral loads. A neutral has different characteristics than an ungrounded conductor. When two ungrounded delta-configured conductors are sharing a neutral, the neutral generally will carry the unbalanced load between phase A and phase B. If phase A has 16 amperes between phase and neutral, and phase B has 10 amperes between phase and neutral, the neutral will carry 6 amperes, or the difference between the two phases. In a delta configuration, the neutral currents are going in opposite directions and cancel each other out. The maximum amperage is subtracted by the remaining amperage, in this case 16 − 10 = 6 amperes on the neutral (see Figure 4-2).

The neutral, when grounded properly, keeps the voltage constant between the phase and ground. If the neutral were not grounded properly, the phase to ground voltage would fluctuate between the phases of a shared neutral. It would act much like a broken neutral. The open neutral is explained in more detail later in this chapter.

A neutral supplied by a wye system is figured differently than a delta neutral. The maximum unbalance is still the largest phase-to-ground load per phase, but when shared by three ungrounded conductors, the following formula will show the amperage on the neutral at any given time:

$$\sqrt{A^2 + B^2 + C^2 - AB - AC - BC}$$

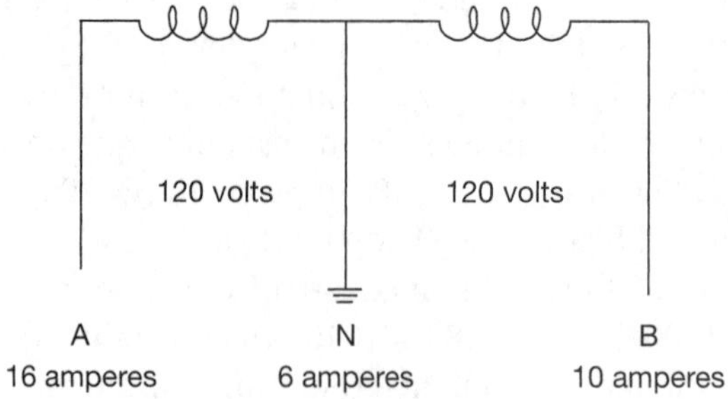

A	N	B
16 amperes	6 amperes	10 amperes

120 volts   120 volts

Figure 4-2

The following phase-to-ground neutral loads were observed: $A = 16$ amperes, $B = 10$ amperes, $C = 5$ amperes, what is the neutral amperage?

$$\sqrt{16^2 + 10^2 + 5^2 - 16 \times 10 - 16 \times 5 - 10 \times 5}$$
$$= \sqrt{256 + 100 + 25 - 160 - 80 - 50}$$
$$= \sqrt{91} = 9.5 \text{ amperes on the neutral.}$$

If two ungrounded conductors derived from a wye share a neutral, there will be amperage on the neutral. A phase has 8 amperes and B phase has 6 amperes. What is the amperage on the neutral?

$$N = \sqrt{A^2 + B^2 - AB} = N = \sqrt{64 + 36 - 48} = \sqrt{100 - 48}$$
$$= \sqrt{52} = 7.211 \text{ amperes}$$

In a wye configuration, the phases are 120 degrees from each other. When two ungrounded conductors share a neutral, the neutral splits the 120 degrees (see Figure 4-3).

If the ungrounded conductors have 10 amperes on each of the two legs, the neutral will have the following amperage:

$$\sqrt{A^2 + B^2 - AB} = \sqrt{100 + 100 - 100} = \sqrt{100} = 10 \text{ amperes}$$

When the third phase is used to share the neutral, it becomes 180 degrees out from the neutral and subtracts ampere for ampere what is already on the neutral from the other two phases.

The following table illustrates what happens to the amperage with two phases constant and the third phase fluctuating between 2 and 12 amperes:

Phase A	Phase B	Neutral
12	12	12

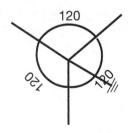

Figure 4-3

When the third phase is added, the amperage changes as the third phase to neutral changes. Once it hits 12 amperes on the third phase, the neutral amperage becomes 0:

Phase A	Phase B	Phase C	Neutral
12	12	2	10
12	12	4	8
12	12	6	6
12	12	8	4
12	12	10	2
12	12	12	0

The above calculations to find the amperage on the neutral used the formula

$$N = \sqrt{A^2 + B^2 + C^2 - AB - AC - BC}$$

The same formula can be used to determine the amperage at any given point. Nonlinear loads, which include fluorescent lighting, have an adverse effect on the neutral. Nonlinear loads are loads derived from electronic ballasts and computer equipment. These loads cause harmonics of different frequencies. These harmful harmonics cause excessive heating on neutral conductors that can cause damage to the conductor. Many commercial installations have compensated for this by running a neutral with each phase and doubling the neutral of the feeder. There are other solutions as well.

The neutral can be reduced for a feeder supplying an oven, cook top, or range to 70% of the load of the ungrounded conductors, as determined by Table 220.55. The neutral can be reduced for a feeder supplying dryers to 70% of the ungrounded conductors, in accordance with Table 220.54.

A further demand factor of 70% is allowed for neutrals over 200 amperes; however, any portion of the load that has a nonlinear load cannot be reduced. No reduction for a neutral over 200 amperes that is supplied by a wye-connected, three-phase, four-wire system, nor a three-wire, two ungrounded and one neutral, of a four-wire three-phase wye system can be derated (Part C, Article 220.61) .

Open Neutral

Sometimes, the neutral is broken and does not have a direct path back to the source. This causes lights to be brighter or dimmer than normal. It

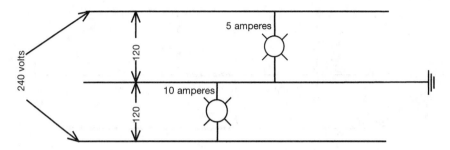

Figure 4-4

can also cause 110 volt motors to operate at too high or too low a voltage. Normally, the circuit is parallel. Voltage is constant, amperage increases as the load increases. A normal three-wire circuit is shown in Figure 4-4. In this circuit, the voltage is constant. The load consists of 5 amperes on one side of the circuit and 10 amperes on the other side. Now, the neutral breaks, removing the solid ground from the circuit. This changes the basic character of the circuit from parallel to series. Instead of the voltage being constant and the amperage changing with the load, the opposite is true. To find the voltage of a circuit with a broken neutral, rules of a series circuit must be applied, as a new path has been established.

An open neutral circuit is shown in Figure 4-5. First, the resistance of each load must be determined and then *all* the resistances are added together: 24 ohms + 12 ohms = 36 ohms.

$$\frac{E}{I} = R \frac{120}{5} = 24 \text{ ohms} \quad \frac{120}{10} = 12 \text{ ohms}$$

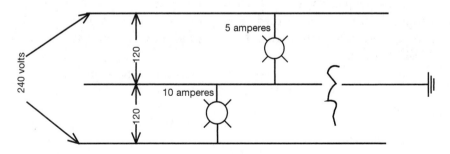

Figure 4-5

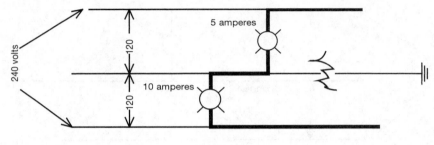

Figure 4-6

Next, the amperage for the circuit must be determined. The total voltage for the circuit is 240 volts:

$$\frac{E}{R} = I \quad \text{or} \quad \frac{240}{36} = 6.66 \text{ amperes}$$

Now, to find the voltage across each load, $I \times R = E$:

$$6.66 \times 24 = 159.84$$
$$6.66 \times 12 = 79.92$$
$$\text{Total voltage} = 239.76$$

In the case of a broken neutral (see Figure 4-6), damage can often result due to voltage too high or too low. The solution to this problem is trace the neutral wire and find and repair the break.

4-10 OVERCURRENT PROTECTION

Overcurrent protection primarily is to protect the conductors, equipment, and people. The general rule for small conductors is that the overcurrent device shall not exceed 15 amperes for #14, 20 amperes for #12, and 30 amperes for #10 conductors unless specifically permitted in Article 240.4 E through G.

Article 240.4 E through G covers specific equipment or conditions that permit larger overcurrent devices to be used on a specific circuit. These include:

- Tap conductors
- Transformer secondary conductors

- Air conditioning and refrigeration
- Capacitor circuits
- Control and Instrumentation circuits
- Electric welder circuits
- Fire alarm circuits
- Motor-operated appliance
- Motor and motor control circuits
- Phase converter supply circuits
- Remote control signaling and power-limited circuits
- Secondary tie conductors

Air Conditioning and Refrigeration Equipment

In general, for air conditioning and refrigeration equipment overcurrent is 175% of full load current. If the overcurrent device is not large enough to handle the starting current, then the overcurrent may be increased to 225%.

Since most air conditioning equipment is rated in tons, it is necessary to convert to equivalent horsepower. To find the equivalent horsepower of a hermetic-refrigerant motor, use the nameplate amperage (full load current) and find the equivalent horsepower in Tables 430.248, 430-249, or 430.250. The equivalent horsepower for locked-rotor current can be obtained from Table 430.251A & B. If the amperage is between horsepower ratings, use the higher rating.

Example: A nameplate rating of a 460 volt hermetic compressor is 45 amperes. What is the equivalent horsepower? In Table 430.250, a 30 horsepower motor is 40 amperes and a 40 horsepower motor is 52 amperes. Since 45 amperes is above the 30 horsepower rating of 40 amperes, but below the 40 horsepower rating of 52 amperes, the equivalent horsepower would be 40 horsepower.

When the motor compressor is the largest load on a branch circuit, the short circuit and ground-fault protective device cannot be larger than specified in Table 440.22B plus the sum of the rated load current or branch circuit protection selection current, whichever is greater, of the other motor compressor and the ratings of the other loads supplied.

When the motor compressor is not the largest load, the rating or setting of the branch circuit short circuit protection and ground fault protective device shall not exceed a value equal to the sum of the rated load current or branch selection current.

Welders

Supply conductors for arc welders are determined by the $I_{1 \text{ eff}}$. If the $I_{1 \text{ eff}}$ value is not given, then it can be determined by multiplying the rated primary current times the factor in Table 630.11A based on the duty cycle of the welder.

Example: A primary current of a nonmotor arc welder is 40 amperes and the duty cycle is 80. What is the amperage used to size the conductors?

$$40 \times .89 = 35.6 \text{ amperes}$$

For a group of welders, the conductor ampacity is determined by Table 630.11A as the sum of 100% of the two largest welders, plus 85% of the third-largest welder, plus 70% of the fourth-largest welder, plus 60% of all of the remaining welders.

Example: Six arc welders have the following currents and duty cycles. Two of the welders have a primary current of 50 amperes, 80% duty cycle; two welders are 40 amperes at 90% duty cycle; and two welders are 30 amperes at 90% duty cycle:

> Welder at 50 amperes $\times$ 0.89 = 44.5 amperes
> Welder at 40 amperes $\times$ 0.95 = 38 amperes
> Welder at 30 amperes $\times$ 0.95 = 28.5 amperes
> Largest two welders = 44.5 $\times$ 2 $\times$ 100% = 89 amperes
> Next-largest welder = 38 $\times$ 1 $\times$ 85%   = 32.3 amperes
> Next-largest welder = 38 $\times$ 1 $\times$ 70%   = 26.6 amperes
> Remaining welders = 28.5 $\times$ 2 $\times$ 60%   = 34.2 amperes
>
> Total   182.1 amperes

The overcurrent device size for the supply conductors for the above group of welders would be 200% of the conductor ampacity:

$$182.1 \times 200\% = 364.2 \text{ amperes}$$

Table 630.12 states that when the amperage value does not correspond to a standard overcurrent device, the rating or setting specified results in unnecessary opening of the overcurrent device and the next-higher standard rating or setting should be used. In this example, a 400 ampere breaker would be appropriate. In the example, each welder must have overcurrent protection of no more than 200% of $I_{1\,max}$, or if $I_{1\,max}$ is not given, then use 200% of the rated primary current of the welder.

Resistance welders are a little different from arc welders. The supply conductors for a welder should not be less than 70% of the rated primary current or duty cycle for seam and automatic feed welders and 50% of the rated primary current for manually operated nonautomatic welders. The ampacity of supply conductors wired for a specific operation when the actual primary current and the duty cycle is known and remain unchanged should not be less than the primary current times the duty cycle multiplier found in Table 630.31A2 for the duty cycle at which the welder will be operated.

Example: For a resistance welder with a primary current of 40 amperes and a duty cycle of 30, the ampacity would be $40 \times 0.55 = 22$ amperes.

For groups of welders, the supply conductors should not be less than the sum of the largest welder supplied according to the values in Table 630.31A2, and 60% of the values obtained for all the other welders supplied.

Example: Five welders all having a primary current of 40 amperes and a duty cycle of 30%:

$$40 \times .55 \times 1 \times 100\% = 22$$
$$40 \times .55 \times 4 \times 60\% = 52.8$$
$$\overline{\text{Total} = 74.8 \text{ amperes}}$$

The overcurrent device cannot be rated at more than 300% of the primary current of the welder. In a group installation, the overcurrent device cannot be set at more than 300% of the conductor rating.

In the above example, 75 = #4, which is good for 85 amperes. The overcurrent device should not exceed 300% of 75, which would be 225 amperes. A 225 ampere breaker would be appropriate. (If the fuse size had been between standard fuse sizes, the next-higher fuse would be acceptable. Standard fuse sizes can be found in Article 240.6.)

Transformers

Overcurrent protection for transformers is divided into two groups: over 600 volts (Table 450.3A) and 600 volts and less (Table 450.3B). In transformers of over 600 volts, the impedance of the transformer must be considered. The impedance is the resistance of the windings of the transformer. The impedance value is on the nameplate. Table 450.3A shows maximum rating of the overcurrent device according to impedance and voltage. Impedance is divided into two groups: not more than 6% and more than 6% but less than 10%. In the table, there is a distinction between circuit breakers and fuses. Secondary protection over 600 volts follows along the same lines, taking into account the impedance and whether fuses or circuit breakers are used. For 600 volts and below, there is no difference between circuit breakers and fuses. Also note that there is no difference in impedance when selecting overcurrent protection for the secondary of 600 volts or less.

Table 450.3B covers the maximum rating or setting of overcurrent protection for transformers of 600 volts and less (as a percentage of transformer rated current). In this instance (transformers of 600 volts and less), impedance is not an issue. Instead, the overcurrent protection is predicated by the amperage: currents of 9 amperes or more, currents of 9 amperes or less, and currents of 2 amperes or less. Each category has a different percentage of amperage to arrive at the proper overcurrent protection. It is further divided between the primary protection and secondary protection. There are notes at the bottom of the table that must be followed when indicated.

Transformers are sized in kVA (1000 × volts × amperes). To find the amperage of a transformer, either the primary or secondary, two formulas are used, one for three phase (voltage × 1.732 × amperes = kVA), and for single phase (voltage × amperes = kVA). If the kVA is divided by the voltage, or in the case of three phase, by voltage × 1.732, the result will yield the full load amperage of the secondary or primary.

Example A: Single-phase transformer with 15 kVA 480 volt primary and 240 volt secondary:

15,000/480 = 31.25 amperes primary

15,000/240 = 62.5 amperes secondary

Example B: Three-phase 15 kVA transformer, 480 volt primary 240 volt secondary:

15,000/480 × 1.732 = 18.04 amperes primary

15,000/240 × 1.732 = 36.08 amperes secondary

The full load amperage is the basis for sizing the overcurrent protection discussed earlier in this chapter. Another point to note is that trans-

Table 4-1. Standard transformer sizes and full load current

Single Phase				Three Phase			
kVA	208	240	480	kVA	208	240	480
1	4.8	4.16	2.08	3	8.3	7.2	3.6
3	14.42	12.5	6.25	6	16.6	14.4	7.2
5	24	20.83	10.42	9	25	21.6	10.8
7.5	36.05	31.25	15.62	15	41.6	36	18
10	48.07	41.67	20.83	30	83.3	72	36
15	72.12	62.9	31.25	45	125	108	54
25	120	104	52	75	208	180	90
37.5	180	156	78	100	277.7	240	120
50	240	208	104	150	416.6	360	180
75	360	312.5	156	225	625	540.8	270
100	480	416.6	208	300	833	721	361

formers are separately derived systems and can be supplied by a three-wire system (delta) and create a four-wire system (grounded delta or wye). Some standard transformer sizes and full load currents are shown in Table 4-1.

CHAPTER 4 TEST

1. What size EMT conduit is needed for 10 THW #10 copper conductors?

2. What size rigid conduit is needed for 6 #12 and 6 #10 THWN conductors?

3. A fixture box has three whips of three wires each, all #12, and three wires entering the box, three wires exiting the box, and two wires passing through the box. What size box is needed?

4. A fixture box contains one whip with three #12. Three #10s splice to the whip then continue with two other pass-through #10s. What is the cubic volume needed for this box?

5. A junction box is to be used as a pull box. The box will be used as an angle pull. It has one 2" conduit and two 1½" conduits entering and leaving the junction box. What is the minimum length and width of this box if all of the conductors are #4 and larger?

6. A 3" conduit enters a box and leaves the box at 90 degrees (angle pull). The cable in the conduit is rated for more than 600 volts and is

1⅜" in diameter. The cable is unshielded. What is the minimum length and width of the box?

7. What is the purpose of the first semiconductor between the conductor and the insulation in a shielded cable?

8. Four 500 MCM THW conductors are in a conduit (three phases and one neutral). What is the size of the grounding-electrode conductor in copper?

9. A 30 horsepower motor, three phase 480 volt, draws 40 amperes. The overcurrent device is a 400 ampere instantaneous breaker. What size is the equipment-grounding conductor?

10. A neutral feeding an apartment complex carrying only the imbalance calculates out to 340 amperes before derating. What is the net amperage of this in order to size the wire?

11. In a three-phase four-wire wye system, (A) phase to neutral was 25 amperes, (B) phase to neutral was 50 amperes, and (C) phase to neutral was 0 amperes. What is the neutral load in amperage?

12. A hermetic air conditioner draws 24 amperes. What is the circuit breaker size?

13. There are three arc welders: 40 amperes with a duty cycle of 80, 30 amperes with a duty cycle of 90, and 50 amperes with a duty cycle of 70. What size conductors would be required to supply these welders?

14. What would be the overcurrent for this group of welders?

15. A 25 kVA transformer rated at less than 600 volts with an impedance of not more than 6% would require secondary protection of how many amperes?

16. A circuit is made up of four 500 MCM THWN copper conductors, one 4/0 THWN copper conductor, and two 2/0 THWN copper conductors. What is the minimum size pvc schedule 40 conduit needed?

17. An EMT nipple less than 24" long is required between a gutter and box for the circuit in problem 16. What is the minimum size needed?

18. The second semiconductor in a shielded high-voltage cable, just under the shielding, serves what purpose?

    A. To ensure a tight bond between the insulation and shield.

    B. To make the cable more expensive.

    C. To provide more insulation value.

    D. None of the above.

19. What size grounding-electrode conductor is required for a 400 ampere service to a copper ground rod?

20. An apartment complex has one service for 25 units. The cooktops are rated at 11 kW each. What is the neutral load after derating?

21. Each of these apartments has a 5 kW dryer. What is the amperage on the service neutral after the demand factor?

22. Six resistance welders are on one feeder. Two welders are 40 amperes at a duty cycle of 50, two welders are 50 amperes at a duty cycle of 40, and two welders are 30 amperes at a duty cycle of 30. What is the conductor size needed for the feeder for these welders?

23. What size overcurrent device is needed for Problem 22?

24. How many #12 THW conductors can be contained in a 1″ EMT conduit?

25. What is the volume allowance for #6 copper conductor?

MOTORS

5-1   INTRODUCTION

Motors perform vital tasks in an electrical system. They can be configured to push, pull, raise, lower, and pump among many other functions. Many of these operations can be done automatically by using a control circuit.

The characteristics of a motor vary by the many different types that can be used for different applications. However, they all develop a locked-rotor current. This is the inrush of current to get the motor started. Once the motor gets up to running speed, the amperage drops down to a full load current. This is equivalent to what happens in a car: a lot of power is needed to get a car from 0 miles per hour to 60 miles per hour. Once the car reaches 60 miles per hour, it takes much less power to maintain the car at 60 miles per hour. In a motor, the amperage needed to start the motor is called starting current or inrush current. It can also be called locked-rotor current.

The inrush current or locked-rotor current on a 460 volt, three-phase, 40 horsepower motor is 290 amperes [see Table 430.251B of the *National Electrical Code®* (NEC)],* whereas the full load current of the same motor is 52 amps (Table 430.250).

---

*This chapter contains material reprinted with permission from NFPA 70-2005, *National Electrical Code®*, Copyright © 2004, National Fire Protection Association, Quincy, MA 02269. This reprinted material is not the complete and official position of the NFPA on the referenced subject, which is represented only by the standard in its entirety. National Electrical Code and NEC are registered trademarks of the National Fire Protection Association, Quincy, MA.

The components of a motor circuit are as follows:

1. Branch circuit conductors
2. Branch circuit short circuit and ground-fault protection
3. Controller
4. Control circuit
5. Running overload protection

5-2 BRANCH CIRCUIT CONDUCTORS

In general, conductors that supply a single continuous duty rated motor shall have an ampacity of a minimum of 125% of full load current.

Example: A 40 horsepower continuous squirrel cage motor at 460 volts draws 40 amperes (Table 430.250): 40 × 1.25 = 50 amperes. The branch circuit conductor size would be #8 THWN, rated for up to 50 amperes (Table 316.16).

Use the values in Tables 430.247, 430.248, 430.249, and 430.250, including all of the notes, to determine the ampacity of the following, instead of the nameplate rating of the motor:

- Ampacity of the conductors
- Rating of switches
- Branch circuit short circuit and ground-fault protection

If the motor is rated in amperes and not horsepower, the equivalent horsepower rating can be found in Table 430.250.

Note that the overloads that are responsive to motor current are selected by the nameplate rating of the motor and not the tables in Article 430 of the NEC.

Motors are rated by duty cycles. Table 430.22E shows the duty cycle service. The left column shows the type of duty for which the motor will be used:

Short-time duty

Intermittent duty

Periodic duty

Varying duty

At the top of the table is the time rating or duty cycle: 5 minute, 15 minute, 30 and 60 minute, and continuous duty. There are different percentages for different time rating and duty cycles. The percentage shown is multiplied by the full load current of the motor involved. The varying of the percentages has to do with the length of operation and the frequency of starting. This can cause different heating characteristics in the conductors and the motor. It should be noted that any motor must be considered continuous duty unless that motor will not operate continuously with a load under any condition.

The conductors on the line side of the controller of a multispeed motor must be based on the current ratings of the windings that the conductors energized.

When the secondary resistor is separate from the controller, the ampacity of the conductors between the controller and the resistor must not be less than that in Table 430-23C.

If there is more than one motor on a feeder, the conductor is calculated by 125% of the largest motor and 100% of the full load currents of the other motors.

Example: A 30 horsepower, 20 horsepower, and 40 horsepower motor are on a single three-phase, 460 volt feeder. What is the size of the conductors of this feeder? (Table 430.250.)

30 horsepower = 40 amperes

20 horsepower = 27 amperes

40 horsepower = 52 amperes

So the largest motor = 52 × 1.25 + 27 + 40 = 132 amperes and the THW conductors would be 1/0 (Table 310-16).

When two motors of equal rating are on a feeder, one of the motors will be considered the largest and calculated at 125%, and the other motor will be added at 100%.

Example: Two 25 horsepower, 460 volt, three-phase motors share the same feeder. What is the ampacity needed to size the conductors? (Table 430-250): 34 × 1.25 + 34 = 76.5 amperes.

If motors of different duties are on a feeder, the amperage is determined by Article 430.22E. The highest amperage calculation is the largest motor and is multiplied by 125%; the rest of the motors are added at 100% of the value of Article 430.22E. This could result in a smaller horsepower motor being the largest motor for this calculation. If a continuous duty motor is in this group, the amperage is found from Tables, 430.248, 430.249, and 430.250.

Example: A 30 horsepower motor rated for short-time duty (30 and 60 minutes) is on a 460 volt, three-phase feeder with a 40 horsepower continuous duty motor. What is the ampacity needed for the feeder conductors?

30 horsepower = 40 amperes × 150% (Table 430.22E) = 60 amperes

40 horsepower = 52 amperes

Since the 30 horsepower motor is the largest load after factoring the duty cycle percentage, it would be multiplied by 125% and the 40 horsepower motor would be added at 100%: 60 × 1.25 + 52 = 127 amperes needed for the feeder.

5-3   MOTOR OVERLOAD PROTECTION

Motors are also protected by overload devices, sometimes called motor heaters. They are designed to protect motors, motor controllers, and motor branch circuit conductors against overheating due to motor overloads and failure to start.

If an overload continues for a long enough period of time, damage or serious overheating could result. Overload devices are different than branch circuit short circuit and ground-fault protection devices (fuses and circuit breakers). Overload devices cannot be substituted for fuses or circuit breakers.

Motors that have a continuous duty rating of either more than one horsepower or less than one horsepower that are automatically started must have overloads based on Table 430.32A1.

If the motor has a service factor of 1.15 or greater or has a temperature rise of 40°C or less, the overloads are 125% of the motor nameplate amperage. All other motors must be calculated at 115% of the motor nameplate amperage. If the overload is not set high enough to start the motor or carry the load, higher-size elements or increment settings can be used, but they cannot exceed 140% of the nameplate rating of the motor if that motor has a marked service factor of 1.15 or greater or a marked temperature rise of 40°C. All other motors must not exceed 130%.

Sometimes, capacitors are used to improve the power factor. The overload device is based on the improved power factor of the motor circuit, but the effect of the capacitor is not considered in calculating the circuit conductors for the motor (Table 460.9).

Some motors are equipped with an approved thermal protector to prevent dangerous overheating of the motor caused by overload and failure to start, but cannot exceed the following percentages of full load current in Tables 430.248, 430.249, and 430.250:

170% for motors 9 amperes or less

156% for motors up to and including 9.1 to 20 amperes

140% for motors of 20 amperes and more

A permanently installed one horsepower motor that is not automatically started and within sight of the controller can be protected by the branch circuit overcurrent device. (Within sight means must be visible and within 50 feet; see Article 100, Definitions.)

The rules for overloads would apply if the motor were not within sight of the controller and the rules for automatically started motors would apply for motors of one horsepower or less (Table 430.32B)

When a motor used for short-time, intermittent, periodic, or varying duty as specified in Table 430.22E it must be protected by the branch circuit short circuit and ground-fault protection device as long as the protective device rating or setting does not exceed that in Table 430.52. A motor shall be considered to be continuous duty unless the process that it drives cannot operate continuously with load under any circumstances. Table 430.37 shows which phases require overload protection.

If the circuit does not exceed 15 amperes at 125 volts or 250 volts, and the motor is one horsepower or less, individual overload protection can be omitted and the motor can be connected by an attachment plug and receptacle. The rating of the attachment plug and receptacle determines the rating of the circuit.

A motor or an appliance over one horsepower must have individual overload protection that is an integral part of the motor or appliance.

If a motor can be restarted after tripping, a motor overload device cannot be installed unless approved by the authority having jurisdiction.

5-4 MOTOR BRANCH CIRCUIT SHORT CIRCUIT AND GROUND FAULT PROTECTION

Fuses and circuit breakers (branch circuit short circuit and ground-fault protection) have two purposes:

1. To protect motor branch circuits, motor control components, and motors against overcurrent caused by short circuits and grounds.
2. They must be able to carry the starting current (locked-rotor current) of the motor.

Table 430.52 shows the maximum setting of the branch circuit short circuit and ground-fault device. It is broken down by the type of motor in the left column and type of overcurrent protection at the top.

If the calculation for the fuse or circuit breaker is in between standard fuse sizes, the next-higher fuse size may be used. Standard size fuses and circuit breakers are shown in Article 240.6.

Example: A 460 volt, three-phase, 30 horsepower motor is to be protected by non-time-delay fuses. From Table 430.250, the amperage of a 460 volt, 30 horsepower motor is found to be 40 amperes. A non-time-delay fuse for a polyphase squirrel cage motor is 300%, so $40 \times 3.00 = 120$ amperes. 120 amperes is not a standard size fuse. The next-highest standard fuse size would be 125 amperes (Table 240.6).

Article 430.52C1, exception 2, allows higher settings for fuses and circuit breakers if the selected fuse or circuit breaker is not adequate to start the motor as follows:

- Non-time-delay fuses of 600 amperes or less or a time delay, class CC fuse up to no more than 400%.
- Time delay fuses can be increased to a maximum of 225%.
- Inverse time circuit breakers to a maximum of 400% for full load currents of 100 amperes or less, or 300% for full load currents of more than 100 amperes.
- A maximum of 300% of a fuse between 601 and 6000 amperes.

Note: If the manufacturer's overload relay table shows the maximum branch circuit short circuit and ground fault for use with a motor controller or it is marked on the equipment, this value must not be exceeded, even if shown in Table 430.52. Only adjustable instantaneous-trip circuit breakers can be used and must be part of a listed combination motor controller having coordinated motor overload and short circuit and ground-fault protection in each conductor, with a maximum value set in Table 430.52. Exceptions are Design B or Design E energy efficient motors that can be set at 800% of the full load current of that motor. This percentage can be raised to 1100% of the full load amperage of Design B and Design E energy efficient motors.

If these values are not adequate to start the motor, a maximum value can be raised to 1300%, except for Design B and Design E energy efficient motors. The values for Design B and Design E energy efficient motors can be set up to a maximum of 1700%.

When two or more motors are on the same feeder and the feeder has been sized at 125% of the largest motor plus the sum of the full load currents of the other motors, the protective device of the branch circuit short circuit and ground fault is sized by the current of the largest motor (Table 430.52) plus the sum of the full load currents of the other motors. If this device does not correspond to a standard fuse or circuit breaker, the next size lower must be used.

In the example of sizing feeder conductors, a 20 horsepower, 30 horsepower, and 40 horsepower 460 volt motors were used to determine the feeder size. Using the same example, the feeder will be protected by a set of non-time-delay fuses. The amperage of the motors is

30 horsepower = 40 amperes

20 horsepower = 27 amperes

40 horsepower = 52 amperes

The largest motor draws 52 amperes. A non-time-delay fuse for this motor would be 300% or 52 × 3 = 156 + 27 + 40 = 223 amperes. The standard size fuse that does not exceed this value would be 200 amperes.

5-5   MOTOR CONTROL CIRCUITS

A motor control circuit carries electrical signals directing the performance of the controller but does not carry the main power circuit. Motor control circuits can be the same voltage as the motor up to 600 volts, or can be reduced by means of a control transformer. Oftentimes a control circuit transformer is used, especially when the control circuit extends beyond the controller. A 460 volt motor with a 120 volt control circuit is much easier to deal with when the circuit extends beyond the controller.

If the control voltage is the same as the motor voltage, then it can be protected by the motor branch circuit short circuit and ground-fault protective device if Table 430.72B is followed. Column B of Table 430.72B shows the maximum setting of the motor branch circuit short circuit and ground-fault device for the size of the control circuit. Notice that #10 copper can be used if the control conductors do not extend beyond the enclosure without additional protection, provided that the branch circuit short circuit and ground fault device are set at 160 amperes or less. If the control circuit extends beyond the controller, the branch circuit short circuit and ground-fault device can only be as high as 90 amperes. If a transformer is used to reduce the voltage of the control circuit, the transformer must be protected in its primary according to Article 430.72(C)(1) through (5).

If one side of the motor control circuit is grounded, the circuit must be used so that an accidental ground will not start the motor or bypass manually operated shutdown devices or automatic safety devices. The best way to avoid doing this is to make sure that the grounded conductor is not used through the control devices such as the start–stop stations. An accidental ground on the grounded conductor can complete the circuit and energize the contactor if used in this manner.

There are several applications of control circuits for a variety of motor applications. Motor contactors are an ideal way to control a motor both manually and automatically. A common way of manually starting and stopping a motor is to use a stop–start push-button station. These stations can be on the controller as well as positioned out in the field. The stop button is a normally closed momentary-contact switch and is in series with the circuit. The start button is a normally open momentary-contact switch and is in parallel with the circuit. When the start button is depressed, the voltage originating at the controller passes through the normally closed stop button. The circuit is complete and energizes a holding coil. Even though the start button returns to the open position immediately after being released, the voltage is maintained through the holding coil and will continue to remain energized until the current to the holding coil is interrupted by pressing the stop button. When the start button is depressed and the holding coil is closed, the magnetic contactor closes in the controller. The motor leads are connected to the bottom of the contactor, whereas the line leads are connected to the top of the contactor. When the magnetic contactor is pulled in and closes, contact is made between the motor leads at the bottom and the line leads at the top, completing the motor circuit, which in turn sends power to the motor. The motor will continue to run until the stop button is depressed, interrupting the circuit (see Figure 5-1).

Another common application is the hand–off–auto switch. This three-position switch has the capability of starting a motor automatically. In the

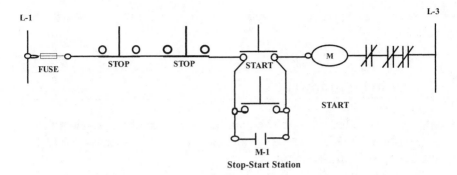

Figure 5-1

off position, the control circuit does not have a path to the holding coil, and the motor will not start. In the "hand" position, a completed path to the holding coil is established and the motor will run continuously until it is turned back to the off position. This position is used to test the motor for rotation and other functions since it bypasses the automatic side. In the automatic mode, a device such as a float switch, pressure switch, or limit switch is connected in series to the control circuit. When this pilot device is closed, the circuit is complete and the motor will run (see Figure 5-2).

An example of this would be a motor running a pump to fill a tank with water. A hand–off–auto circuit is used to run the motor. A float switch is set in the tank and wired in series to the automatic side of the circuit. When the water is low, the contacts in the float switch close, completing the circuit and allowing the motor to start filling the tank with water. When the tank has been filled, the float switch rises and the

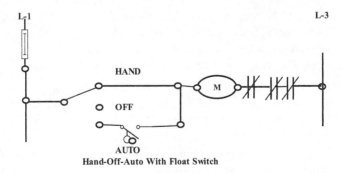

Figure 5-2

contacts open, thereby interrupting the circuit, and the motor will stop running.

5-6   MOTOR CONTROLLERS

A motor controller can be a variety of devices. It can be any switch or device that is normally used to start and stop a motor by making and breaking the motor control circuit.

The branch circuit short circuit and ground-fault device can be used as the controller for stationary motors of ⅛ horsepower or less that are normally left running and cannot be damaged by overload or failure to start. A good example of this would be a clock motor.

An attachment plug and receptacle can be used as a controller for a portable motor of ⅓ horsepower or less.

A controller must be able of starting and stopping the motor it controls as well as able to interrupt the locked rotor current of the motor.

Unless an inverse-time circuit breaker or molded case switch is used, controllers must have horsepower rating at the applied voltage.

If a Design E motor of 3 to 100 horsepower is being used, the controller must be rated for that use or must have a horsepower rating not less than 1.4 times the rating of the motor; if over 100 horsepower, it must have not less than 1.3 times the rating. An inverse-time circuit breaker or molded case switch rated in amperes can be used as a controller for all motors.

For motors of 2 horsepower or less, a general use switch must have twice the amperage rating of the motor. A general use snap switch can only be used in alternating circuits. A motor's amperage must be 80% or less of the rating of the switch. A general use ac–dc snap switch is not suitable.

Controllers must be rated for continuous duty for torque motors, the full load current rating to be not less than the nameplate rating of the motor. If the controller is rated in horsepower but not in amperes, the equivalent current rating can be calculated by finding the horsepower rating form Tables 430.247. 430.248, 430.249, and 430.250.

Unless the motor controller is also the disconnecting means, it is not required to open all conductors. If all the conductors of the circuit can be opened at the same time, then one pole of the controller can be permanently grounded.

Individual controllers must be provided for each motor unless the motor is under 600 volts, or there is a single machine with several motors, or a single overcurrent device, or a group of motors located in a single room within sight of the controller.

5-7 DISCONNECTING MEANS

All ungrounded conductors must be opened by the disconnecting means; no pole can operate independently. The disconnecting means can be in the same enclosure as the controller.

All disconnecting switches must plainly indicate whether they are open (open) or closed (on).

For motor circuits rated at 600 volts or less, the disconnecting means must be at least 115% of the full load current rating of the motor. The disconnecting means can be branch circuit fuses or circuit breakers. For stationary motors rated more than 40 horsepower dc or 100 horsepower ac, a general use or isolating switch can be used but must be plainly marked "Do not operate under load."

The disconnecting means other than the branch circuit short circuit and ground-fault protective device is used as a safety device to disconnect the motor circuit. It should be in sight of the motor and can be unfused. If a person is working on the motor, the disconnect will be where he or she can see it; that protects the person from a motor accidentally starting. If the branch circuit short circuit and ground-fault protective device is used as a disconnecting means and is not within sight of the motor, it must be capable of being locked in the open position.

For combination loads with a motor or motors and other loads that can be started simultaneously on one disconnecting means, the combined ampere and horsepower is determined as follows:

1. Add all of the loads, motor loads, and resistance loads at 100%. This will yield the equivalent horsepower.
2. Add the locked-rotor current in amperes to the other load(s) to obtain an equivalent locked-rotor current for the combined load.

Example: A 30 horsepower and a 40 horsepower 460 volt motor along with a 10 kW heat strip will be under a single disconnecting means. What size disconnect is needed?

First, add all of the currents at full load, then add the locked-rotor currents.

Full load currents		Locked rotor currents
30 horsepower	= 40 amperes	218 amperes
40 horsepower	= 52 amperes	290 amperes
10 kW heat strip	= 12 amperes	12 amperes
	104 amperes	520 amperes

104 amperes at full load current equivalent horsepower = 100 horsepower. 520 amperes at locked rotor current equivalent horsepower = 75 horsepower. The disconnecting means is 115% of 104 amperes or 119.6 amperes. The minimum switch size would be 125 amperes or a 100 horsepower switch.

Summary

1. Motor branch circuit conductors, 125% (Table 430.22)
2. Motor branch circuit short circuit and ground-fault protection (Table 430.52), 150%–700%
3. Overload protection (Table 430.32), 115%–125%
4. Control circuit protection, (Table 430.72B)
5. Disconnecting means (Table 430.101), 115%

Example: For a 40 horsepower, three-phase, 460 volt motor, calculate the motor branch circuit conductors (THWN), conduit size (EMT), fuse size (non-time-delay), overload protection (motor is marked for temperature rise of 40 degrees Celsius or less), and a nonfused disconnecting means at the motor.

A 40 horsepower, three-phase, 460 volt motor draws 52 amperes (Table 430.250). Then

1. Branch circuit size: $52 \times 1.25 = 65$ amperes for three #6 THWN conductors (Table 310.16). Equipment ground: one #6 (Table 250.122). Conduit size in EMT: four #6 = 1¼ inch (Annex C, Table C1).
2. Fuse size: non-time-delay, $52 \times 300\% = 156$ amperes; next standard fuse size up = 175 amperes (Table 240.6).
3. Overload protection responsive to motor current: $52 \times 1.15 = 59.8$ amperes.
4. Control circuit stop–start. Since the motor branch circuit short circuit

and ground-fault protection exceeds the values in Table 430.72B for #10 and #12 copper conductors, supplementary overcurrent protection would be required for the control circuit inside and outside of the controller enclosure.

5. Disconnecting means at the motor (nonfused switch): $52 \times 1.15 = 59.8$ amperes. Standard size switch = 60 amperes.

5-8   FIRE PUMPS, CONTROLLERS, AND MOTORS

Fire pumps, controllers, and motors are calculated differently from nearly all other motors. Motor circuits, in general, are designed to protect the motor, controller, branch circuit conductors, and all of the associated equipment from damage as well as to protect people.

In the case of fire pumps, controllers, and motors, the code does not try to protect the equipment. The intention, rather, is to fight the fire until the last possible second. The conductors are 125% of the full load current of the motor. They must not have a rating less than 125% of the sum of the fire pump motor(s) and pressure maintenance motor(s) at full load currents, and 100% of the associated fire pump accessory equipment.

The overcurrent device must be able to carry the locked-rotor current of the pump motor(s) and the pressure maintenance pump motor(s) and the full load current of all of the fire pump equipment, indefinitely. Secondary overcurrent protection is not permitted.

Power circuits must not have automatic overload protection. Branch circuits are protected against short circuit protection only.

Feeders for fire fighting equipment are supplied separately from the main service or are tapped ahead of the service. The disconnecting means must be able to be locked in the closed position.

CHAPTER 5 TEST

1. What is locked-rotor current?
2. What is full load current?
3. A fire pump draws a full load current of 28 amperes and a locked-rotor current of 162 amperes. What is the correct amperage to determine the overcurrent device?
4. What is the amperage to determine wire size for Problem 3?

5. A 40 horsepower wound-rotor motor, three-phase at 460 volts, requires what size THWN copper conductors?

6. Two 30 horsepower motors and a 40 horsepower motor, all 460 volt, three-phase, continuous duty squirrel cage motors, are on a single feeder. Size the feeder in THW copper conductors.

7. What size overcurrent protection is required for Problem 5 if non-time-delay fuses are used?

8. What size dual-element fuses would be required for Problem 6?

9. Overloads in a motor circuit do not provide what?

10. A 40 horsepower wound-rotor motor at 460 volts, three-phase is rated for varying duty at 30 and 60 minute ratings. What is the ampacity of the conductors?

11. What size overloads are required for a 10 horsepower motor at 460 volts, three-phase, with a 1.15 service factor?

12. An instantaneous breaker is required for a Design E motor. What is the maximum setting for this breaker in amperes according to Table 430.52?

13. A branch circuit short circuit and ground-fault device can protect a control circuit of #10 copper up to a maximum of _____ amperes.

14. What is the maximum amperage for the #10 if the control circuit extends beyond the enclosure and is still be protected by the branch circuit short circuit and ground-fault protection?

15. In a stop–start control circuit, the stop is in _____ and the start is in _____.

16. What keeps the circuit closed when the start button is released?

17. In a hand–off–automatic control circuit, a device is in series on the automatic side of the circuit, which opens and closes the circuit, thereby controlling the motor stopping and starting. True or false?

18. What provisions must be made if the branch circuit short circuit and ground-fault protective device is used as the disconnecting means and is not within sight of the motor?

19. What size disconnecting means would be required for a 460 volt, three-phase, 40 horsepower motor?

20. Isolating switches for stationary motors rated at more than 40 horsepower dc or 100 horsepower ac are permitted to be general use or isolating switches, but must be plainly marked "__ ___ _____ ____ _____."

21. A 40 horsepower motor is on the same feeder as a 20 kW heater. Both are 460 volt, three-phase. What is the equivalent horsepower?

22. What size disconnect is required for Problem 21?

23. A 230 volt, three-phase, 20 horsepower motor has a service factor of 1.15 and a temperature rise of 40°C. What size overloads are required?

24. An elevator motor is rated as continuous duty. What is the percentage of ampacity used to calculate the conductors?

25. What can be used as a controller for a stationary motor of ⅛ horsepower or less that is normally left running and cannot be damaged by overload or failure to start?

chapter 6

SINGLE-DWELLING
LOAD CALCULATIONS

6-1 GENERAL LIGHTING AND RECEPTACLES

Lighting calculations are based on the outside dimensions of the dwelling expressed in square feet. Portions of the house, such as open porches, garages, or unused unfinished spaces, are not included in the dimensions.

Once the floor space or dimensions have been determined, the watts per square foot can be used [see Table 220.12 of the *Natinal Electrical Code* (NEC)].* Dwelling units are 3 watts per square foot. In dwelling occupancies, general-use receptacles rated at 20 amperes or less are included in the lighting load of 3 watts per square foot. No additional calculation is required for general-use receptacles.

Example: The outside dimensions of a dwelling are 2400 square feet. What is the gross lighting and receptacle load?

$$2400 \times 3 = 7200 \text{ watts}$$

To calculate how many circuits are required for the general lighting and receptacles, divide the gross lighting watts by 120 volts, then by either 15 or 20 amperes to get the number of circuits needed:

*This chapter contains material reprinted with permission from NFPA 70-2005, *National Electrical Code®*, Copyright © 2004, National Fire Protection Association, Quincy, MA 02269. This reprinted material is not the complete and official position of the NFPA on the referenced subject, which is represented only by the standard in its entirety. National Electrical Code and NEC are registered trademarks of the National Fire Protection Association, Quincy, MA.

$$7200/120 = 60/15 = 4 - 15 \text{ ampere circuits}$$

$$60/20 = 3 - 20 \text{ ampere circuits}$$

An additional 20 ampere circuit for the bathroom is required, but no additional calculation is needed since it is included in the 3 watts per square foot.

Two small-appliance circuits are required in the kitchen, pantry, dining room, or similar area. Each circuit should be computed at 1500 watts each (Articles 210.11C1 and 220.52A).

In addition, a laundry circuit is required. This circuit is computed at 1500 watts (Tables 210.11C2 and 220.52B).

The lighting and receptacle load, two small-appliance circuits, and laundry circuit can be added together:

```
7200 watts per square foot lighting and receptacle load
3000 watts for two 1500 watt small-appliance circuits
1500 watts for a 1500 watt laundry circuit
11,700 watts subtotal
```

The 11,700 watts can now be derated using Table 220.11. For dwelling occupancies, the first 3000 watts is calculated at 100%. 3001 to 120,000 watts is calculated at 35%. If there is over 120,000 watts, the remainder is calculated at 25%. This same table is also used for multifamily dwellings, which will be covered in the next chapter.

```
11,700 watts subtotal
- 3000 watts × 100% = 3000 watts
  8700 watts × 35%  = 3045 watts
                      6045 watts
```

The 6045 watts obtained is used to determine the size of the feeder or service. It is not used to determine the number of branch circuits required. That was done in the earlier example.

6-2   OVENS AND COOKTOPS

Ovens and cooktops are covered in Table 220.55. For ranges over 8¾ and up to 12 kW the demand factor brings the range down to 8 kW. If the range exceeds 12 kW, for example, a 15 kW range, the first 12 kW is calculated at 8 kW. Each kilowatt or major fraction above 12 kW increases the 8 kW by 5%:

$$
\begin{aligned}
&15 \text{ kW range} \\
&\underline{-12 \text{ kW}} = 8 \text{ kW} \\
&3 \times 5 = \underline{15\%} \\
&1.2 \text{ kW} + 8 \text{ kW} = 9.2 \text{ kW}
\end{aligned}
$$

9200 watts would be the calculation for the service. Since this is one range, the branch circuit can also be determined on the Table 220.55 as 9200 watts. 9200/230 = 40 amperes. The neutral for this range is calculated at 70% of the calculated load for the feeder or service: 9200 × 0.70 = 6440 watts.

The notes in Table 220.55 show the different combinations of cooking equipment and how to calculate the branch circuit conductors as well as the feeders. To calculate the branch circuit conductors for one counter-mounted cooktop unit and one wall-mounted oven, add the nameplate rating to the note for the appliance in Table 220.55. Another example from the same note in the table shows that a counter-mounted cook top and no more than two wall-mounted ovens all supplied from a single branch circuit and in the same room can have their nameplate ratings added together and be treated as a single range; then they can be derated using Table 220.55.

Example:

$$
\begin{aligned}
&2 \text{ wall-mounted ovens, 6 kW each} \\
&1 \text{ counter-mounted cooktop unit, 4 kW} \\
&6 \times 2 = 12 + 4 = 16 \text{ kW}
\end{aligned}
$$

$$
\begin{aligned}
&16 \text{ kW} \\
&\underline{-12 \text{ kW}} = 8 \text{ kW} \\
&4 \text{ kW} \times 5 = 20\% \times \underline{1.20} \\
&\phantom{-14 \text{ kW} \times 5 = 20\% \times }9.6 \text{ kW or } 9.6 \times 1000 = 9600 \text{ watts}
\end{aligned}
$$

The neutral would be 70% of the computed load: 9600 × 0.70 = 6720 watts.

If a 3 kW cooktop unit and an 8 kW oven were used:

$$
\begin{aligned}
&3 \text{ kW cooktop (column B, Table 220.19)} \times 0.80 = 2.4 \text{ kW} \\
&8 \text{ kW oven (column C, Table 220.19)} \times 0.80 = \underline{6.4} \text{ kW} \\
&\text{Total} = 8.8 \text{ kW}
\end{aligned}
$$

6-3   HEATING AND AIR CONDITIONING

When two different loads are on the feeder and will not be used at the same time, the smaller of the two loads can be omitted when calculating the total load for the feeder or service (see NEC, Article 220.60).

Since air conditioning and heating are not likely to be on at the same time, the smaller of the two loads may be omitted when determining the feeder or service.

Generally, heating is rated in kilowatts and air conditioning is rated in amperes. In order to calculate which load is greater, one of the loads must be converted from amperes to kilowatts or from kilowatts to amperes.

Example: An air conditioner draws 24 amperes and the heating is 10 kW; which is larger?

$$24 \times 240 \qquad = \quad 5760 \text{ watts}$$
$$10 \text{ kW} \times 1000 = 10{,}000 \text{ watts}$$

Since the 10 kW heating exceeds the 5.8 kW air conditioning, the heating is calculated into the feeder or service and the air conditioning is omitted.

6-4   ELECTRIC CLOTHES DRYER

The minimum value that can be used for a household electric clothes dryer is 5000 watts or the nameplate rating of the dryer, whichever is greater (Article 220.54).

If more than four electric clothes dryers are used on one service, the demand factor in Article 220.54 can be used. The neutral for an electric clothes dryer is computed at 70% of the computed load (Article 220.61).

Example: 5000 watt electric clothes dryer = 5000 watts. To compute the neutral load for this dryer, 5000 × 0.70 = 3500 watts.

6-5   FIXED APPLIANCES

If there are four or more fixed appliances that are fastened in place, such as water heaters, dishwashers, garbage disposals, and attic fans, a demand factor of 75% can be applied to these appliances on the same feeder or service. Electric ranges, clothes dryers, space-heating equipment, or air conditioning is not included in this demand factor (Article 220.53).

Example: The following fixed appliances are on a single feeder:

1 Dishwasher	600 watts
1 Garbage disposal	800 watts
1 Water heater	2000 watts
2 Attic fans, 500 watts each	1000 watts
	4400 watts × 0.75 = 3300 watts

6-6   LOAD CALCULATION FOR A SINGLE-FAMILY DWELLING

A 2400 square foot single-family dwelling contains the following:

1 14 kW oven

1 central heating unit at 10 kW, 240 volts

1 air conditioner at 6440 watts, 240 volts

1 dishwasher at 600 watts, 120 volts

1 garbage disposal at 800 watts, 120 volts

2 attic fans at 250 watts each, 120 volts

1 clothes dryer at 5000 watts, 240 volts

1 water heater at 2000 watts, 240 volts

Calculate the amperage of the phases and the neutral, then determine the wire size (see Table 6-1, next page).

After the amperage has been determined, the wire size can be found in Table 310.15B6, which allows a further reduction in wire size. This table is only for single-family dwellings.

An example single-family standard calculation for a 1500 square foot single-family dwelling containing

1 cooktop, 14 kw

1 clothes dryer, 4500 watts

1 dishwasher 600 watts, 120 volts

1 garbage disposal 750 watts, 120 volts

1 water heater 4500 watts, 240 volts

6 space heaters, 500 watts each, 240 volts

1 air conditioner 13.6 amperes, 240 volts

is shown in Table 6-2 (page 115).

Table 6-1   Residential load calculation, Example 1

			Phase A	Phase B	Neutral
Lighting: 3 watts per sq ft	2400 × 3 =	7200			
Appliance circuits	1500 × 2 =	3000			
Laundry circuit	1500 × 1 =	1500			
Subtotal		11,700			
First 3000 watts at 100%		− 3000	3000	3000	3000
Remainder at 35% up to		8700 × 0.35	3045	3045	3045
120,000 watts					
First 12 kW at 8 kW	14				
5% of each kW in excess	−12	8000			
of 12 kW	2	× 1.10			
70% of computed load for		8800	8800	8800	
neutral		8800 × 0.7			6160
Heat, 240 volts		10,000	10,000	10,000	
AC, 240 volts		6440 (omit)			
Clothes dryer		5000		5000	5000
70% of computed load for		5000 × 0.7			3500
neutral					
1 Dishwasher, 120 volts		600 × 0.75	450		450
1 Garbage disposal		800 × 0.75		600	600
2 Attic fans, 120 volts		500 × 0.75	375		375
1 Water heater, 240 volts		2000 × 0.75	1500	1500	
Totals:			32,170	31,945	17,130
To find amperage, divide each leg by 240 volts			÷ 240	÷ 240	÷ 240
			136	135	67
To find wire size, go to NEC Table 310-15b6		Copper	#1	#1	#4
		Aluminum	2/0	2/0	#3

6-7   ADDITION TO EXISTING DWELLING

When adding to a dwelling, it must be determined if the existing service size is large enough to handle the addition. Article 220.83A describes an addition where additional air conditioning and or heating is not to be installed.

The following loads must be included:

- 3 watts per square foot
- 2 small appliance circuits at 1500 watts each
- 1 laundry circuit at 1500 watts each

Table 6-2   Residential load calculation, Example 2

			Phase A	Phase B	Neutral
Lighting: 3 watts per sq ft	1500 × 3 =	4500			
Appliance circuits	1500 × 2 =	3000			
Laundry circuit	1500 × 1 =	1500			
Subtotal		9000			
First 3000 watts at 100%		− 3000	3000	3000	3000
Remainder at 35% up to 120,000 watts		6000 × 0.35	2100	2100	2100
Cooktop [a]	3000 × 0.8	2400			
2 Ovens, 6 kW each[b]	12,000 × 0.65	7800			
of 12 kW		10,200	8800	8800	
70% of computed load for neutral		10,200 × 0.7			7140
Heat, 240 volts	3000	3000 (omit)			
AC, 240 volts 13.6 amperes	3264 × 1.25[e]	4080	4080	4080	
Clothes dryer[c]	4500	5000	5000	5000	
70% of computed load for neutral		5000 × 0.7			3500
Fixed appliances[d]					
1 Dishwasher, 120 volts		600	600		600
1 Garbage disposal		750		750	750
1 Water heater, 240 volts		4500	4500	4500	
Totals:			28,080	28,230	17,090
To find amperage, divide each leg by 240 volts			÷ 240	÷ 240	÷ 240
			121	122	75
To find wire size, go to NEC Table 310-15b6	Copper		#2	#2	#4
	Aluminum		1/0	1/0	#3

[a]Cooktop, NEC Table 220-55 Column A, 80% for 1 unit, Note 3.
[b]Ovens, Table 220-55 Column B, 65% for 2 units, Note 3.
[c]Minimum allowed by NEC Article 220 54, 5000 watts.
[d]Larget motor, 125%.

- Electric cooking equipment at nameplate rating
- All other appliances permanently connected to a dedicated circuit at nameplate rating. This includes electric clothes dryers, water heaters, dishwashers, garbage disposals, etc.

The first 8 kW of load is calculated at 100%. The remainder of the load is calculated at 40%, then added together.

Example: A 500 square foot addition is added to the dwelling. Is the service sufficient enough to handle the addition? Note that no air conditioning or heat will be added to this addition.

3 watts per square foot 2400 + 500 = 2900 × 3	=	8700 watts
2 small appliance circuits at 1500 watts each × 2	=	3000 watts
1 laundry circuit at 1500 watts	=	1500 watts
1 range at 14,000 watts	=	14,000 watts
1 dishwasher at 600 watts	=	600 watts
1 garbage disposal at 800 watts	=	800 watts
2 attic fans at 250 watts each	=	500 watts
1 electric water heater at 2000 watts	=	2000 watts
1 electric clothes dryer at 5000 watts	=	5000 watts
1 heater at 10,000 watts	=	10,000 watts
		46,100 watts

$$46,100 \text{ watts}$$
$$-8000 \text{ watts} \times 100\% = 8000$$
$$38,100 \text{ watts} \times 40\% = 15,240$$
$$23,240 \text{ watts} \div 240 = 97 \text{ amperes}$$

The 500 square foot addition was added to the original 2400 square footage. The existing service is #1 copper, which is good for 150 amperes. Therefore the existing service is sufficient to handle the 500 square foot addition.

Article 220.83B is the same as 220.83A except that air conditioning and or heating is allowed to be added to the new addition. To add one 6 ampere, 240 volt air conditioning unit to this room, provide for the additional equipment to be added to new addition as follows:

Air conditioning equipment, 100%

Central electric space heating, 100%

Less than four separately controlled space heating units, 100%

First 8 kW of other load, 100%

Remainder of other load, 40%

The other load consists of the items in the previous example. Since the calculations have already been done in the previous example, add the air conditioning unit at 100% to it:

$$\begin{array}{ll}
\text{Gross wattage} & = 46{,}100 \text{ watts} \\
- 8000 \text{ watts} \times 100\% = & 8000 \text{ watts} \\
38{,}100 \text{ watts at } 40\% & = \underline{15{,}240 \text{ watts}} \\
& 23{,}240 \text{ watts}
\end{array}$$

Now add the air conditioner to the calculation:

$$\begin{array}{ll}
6 \times 240 = & 1380 \text{ watts} \\
+ & \underline{23{,}240 \text{ watts}} \\
& 24{,}620 \text{ watts}
\end{array}$$

$24{,}620 \div 240 = 103$ amperes. Therefore, the original service of 150 amperes is sufficient to handle the new load including the air conditioner.

6-8 OPTIONAL CALCULATION FOR SINGLE-FAMILY DWELLING

Single-family dwellings can also be calculated by the optional method as outlined in Article 220.82. The difference in the calculation methods is that most of the loads are added together at full nameplate rating, then a demand factor is used, rather than using several demand factors as the calculation progresses as in the case of the standard method. The minimum service to a dwelling calculated under the optional method is 100 amperes. The calculation is divided into two parts, B and C. Part C is the air conditioning and heating, and Part B is all the other loads.

Part B:

- 3 watts per square foot
- 2 small appliance circuits at 1500 watts each
- 1 laundry circuit at 1500 watts
- Cooktops, ovens, clothes dryers, water heaters, and all appliances that are permanently connected
- All low-power-factor loads at nameplate rating

The loads above are added together and derated. The first 10 kW is at 100%. The remainder of the load is multiplied by 40%. This calculation is set aside until Part C is computed; then Part B and Part C are added together.

Part C consists of heating and air conditioning, whichever is largest. The smaller of the two loads is omitted. To decide which load is larger, use the following information to make that determination:

- 100% of the air conditioning (nameplate rating)
- Heat pumps and supplemental heating at 100% (nameplate rating) if the controller does not prevent the compressor and supplemental heating from operating at the same time
- The calculation value of 100% of the nameplate rating for electric thermal storage and other heating systems that are expected to be continuous
- 65% of the nameplate rating of the central electric space heating system, including supplemental heating in heat pumps; the controller prevents the compressor and supplemental heating from operating at the same time
- If there are less than four separately controlled units, electric space heating is calculated at 65% of the nameplate rating.
- If there are four or more separately controlled units, use 40% of the nameplate rating of electric space heating.

Use the appropriate heating or air conditioning from Part C and add that to your calculation.

Example: Optional calculation for single-family dwelling: A 2400 square foot house has the following in it (see Table 6-3):

1 oven (14 kW)

1 dishwasher (600 watts)

1 garbage disposal (800 watts)

2 attic fans (250 watts each),

1 water heater (2500 watts)

1 clothes dryer (5000 watts)

central space heating with less than four control units (10,000 watts).

The total gross wattage in the above example is 35,100 watts, excluding electric heat. To finish the calculation, you must take the first 10 kW at 100% and the remainder at 40%:

$$
\begin{array}{rl}
34{,}600 \text{ watts} & \\
-10{,}000 \times 100\% = & 10{,}000 \\
24{,}600 \times 40\% \ \ = & +\,9840 \\
\hline
& 19{,}840
\end{array}
$$

Table 6-3   Residential load calculation, optional Example 3

Lighting: 3 watts per sq ft	$2400 \times 3 =$	7200
Appliance circuits	$1500 \times 2 =$	3000
Laundry circuit	$1500 \times 1 =$	1500
Oven	14,000	14,000
Clothes dryer	5000	5000
Dishwasher	600	600
Garbage disposal	800	800
2 Attic fans, 250 watts each	500	500
Water heater	2000	2000
Total		34,600
	$-10,000 \times 100\% =$	10,000
	$24,600 \times 40\% =$	9840
Subtotal		19,840

The total gross wattage in the above example is 34,600 watts, excluding electric heat. The first 10 kW is at 100% and the remainder at 40%. This gives a subtotal of 19,840 watts. Next, the central electric space heat is calculated at 65% (NEC Table 220.82B5) and is added to the subtotal of 19,840 watts:

Subtotal	19,840	19,840
Electric space heat	$10,000 \times 0.65 =$	6500
		$26,340/240 = 110$ amperes

Next, add in the heat (10 kW of central space heat):

$$10,000 \times 65\% = \frac{+\ 6500}{26,340}$$

26,340 watts ÷ 240 = 110 amperes.

This determines the ungrounded phase conductors only. To determine the neutral conductor, you must go back through and list the phase-to-neutral loads, as in the standard method. To determine the neutral size, make a list of the neutral loads and derate them as in the standard calculation:

3 watts per square foot	$2400 \times 3 =$	7200
2 appliance circuits at 1500 watts each	$1500 \times 2 =$	3000
1 laundry circuit at 1500 watts	$=$	1500
Subtotal	$=$	11,700

1 oven at 14 kW

− 12 kW = 8000 watts	− 3000 × 100% = 3000
2 × 5% = + 10%	8700 × 35% = 3045
8800 × 70%	= 6160
1 Clothes dryer, 5000 watts × 70%	= 3500
1 Dishwasher, 600 watts × 75%	= 480
1 Garbage disposal, 800 watts × 75%	= 600
2 Attic fans, 250 watts each × 75%	= 375
	17,130

17130 ÷ 240 = 71 amperes.

Example: Optional calculation of single-family dwelling, 1500 square foot with the following (see Table 6-4):

1 cooktop, 3 kW, 240 volts

2 ovens, 6 kW each, 240 volts

1 clothes dryer, 4500 watts, 240 volts

1 dishwasher, 600 watts, 120 volts

1 garbage disposal, 750 watts, 120 volts

1 water heater, 4500 watts, 240 volts

6 space heaters, 500 watts each, 120 volts

1 air conditioner, 13.6 amperes, 240 volts

Table 6-4   Residential load calculation, optional Example 4

Lighting: 3 watts per sq ft	1500 × 3	=	4500
Appliance circuits	1500 × 2	=	3000
Laundry circuit	1500 × 1	=	1500
Cooktop	3000 × 1		3000
Oven	6000 × 2	=	12,000
Clothes dryer	5000		5000
Dishwasher	600		600
Garbage disposal	750		750
Water heater	4500		4500
Total			34,850
		− 10,000 × 100% =	10,000
		24,8500 × 40% =	9940
Subtotal			19,940
Subtotal	19,940		19,940
AC 13.6 amps 240 volts	3264 × 100% =		3264
		23,204 ÷ 240 = 97 amperes	

The gross wattage was 34,850 watts in Table 6-4, excluding heat or air conditioning. The first 10 kW is calculated at 100% with the remainder at 40%:

$$34,850$$
$$- 10,000 \times 100\% = 10,000 \text{ watts}$$
$$24,850 \times 40\% = \underline{9940} \text{ watts}$$
$$19,940 \text{ watts}$$

The air conditioning is 3264 watts (13.6 amperes × 240 volts) compared to six space heaters at 500 watts each calculated at 40% (6 × 500 × 40% = 1200 watts) (NEC Articles 220.82C1 and 6). The heating is dropped since it is unlikely that it will be on at the same time as the air conditioning. The air conditioning load is added directly to the net calculation of 19,940 watts:

$$19,940 \text{ watts}$$
$$+ 3264$$
$$23,204 \text{ watts} \div 240 = 97 \text{ amperes}$$

The above calculation determined the ungrounded conductors only. To determine the neutral, the standard calculation must be used (see Table 6-5).

Table 6-5   Calculation to determine the neutral

Lighting	4500		
Appliance circuits	3000		
Laundry circuit	1500		
	9000		
	$- 3000 \times 100\% = 3000$		
	$6000 \times 35\% = 2100$		
	5100		5100
Oven	$3 \text{ kW}^a \times 0.8 = 2400 \times 0.7$	$= 1680$	
	$12 \text{ kW}^b \times 0.65 \quad 7800 \times 0.7$	$= 7140$	
	10,200		
Heat^c	3000		3000
Dryer^d	$5000 \times 0.7 =$		3500
Dishwasher	600		600
Garbage disposal	750		750
Total			$21,770 \div 240 = 91 \text{ amperes}$

^{a}Cooktop, NEC Table 220.55, Column A, 80% for 1 unit, Note 3.
^{b}Ovens, Table 220-55, Column B, 65% for 2 units, Note 3.
^{c}Even though the air conditioning was larger, the heat is a neutral load and must be used.

6-9 OPTIONAL CALCULATION FOR A TWO-DWELLING UNIT

Single-family dwellings are single-family houses; multifamily dwellings include apartments, condos, and townhouses. A multifamily dwelling is by definition a building that contains three or more dwelling units. In between the single-family dwelling and the multifamily dwelling is the two-family dwelling. For a service feeding a two-family dwelling, Article 220.85 states that if the calculation under the standard method of the two units exceeds that of three identical units computed under the optional method, the lesser of the two loads can be used.

Example: A 1200 square feet, two-family dwelling has the following in each unit:

Table 6-6   Standard method-two units

			Phase A	Phase B	Neutral
Lighting: 3 watts per sq ft	$1200 \times 3 \times 2 =$	7200			
Appliance circuits	$1500 \times 2 \times 2 =$	6000			
Laundry circuit	$1500 \times 1 \times 2 =$	3000			
Subtotal		16,200			
First 3000 watts at 100%		-3000	3000	3000	3000
Remainder at 35% up to		$13,200 \times 0.35$	4620	4620	4620
120,000 watts					
Oven[a]	$8 \text{ kW} \times 2 \times 0.65$		104,000	104,000	
70% of computed load for neutral	10,400	$\times 0.7$			7280
Heat, 240 volt	750×3		4500	4500	
Clothes dryer	5000×2		10,000	10,000	
70% of computed load for neutral		$10,000 \times 0.7$			7000
1 Dishwasher 120 volts[b]	600×2	$\times 0.75$	900		900
1 Garbage disposal 120 volts[b]	750×2	$\times 0.75$		1125	1125
1 Water heater, 240 volts[b]	2500×2	$\times 0.75$	3750	3750	
Totals:			37,170	37,395	23,925
			$\div 240$	$\div 240$	$\div 240$
			$= 155$	$= 156$	$= 100$

[a]If oven is less than 8¾ kW, column B of NEC Table 220.55 applies.
[b]There are three fixed appliances in each unit; however, when combined in one service, there are six fixed appliances.

Table 6-7 Optional method, three units

Lighting:	$1200 \times 3 \times 3 = 10{,}800$
Appliance circuits	$3000 \times 1 \times 3 = 9000$
Laundry circuits	$1500 \times 1 \times 3 = 4500$
Oven	$8000 \times 3 = 24{,}000$
Clothes dryer	$5000 \times 3 = 15{,}000$
Dishwasher	$600 \times 3 = 1800$
Garbage disposal	$750 \times 3 = 2250$
Water heater	$2500 \times 3 = 7500$
Heat	$2250 \times 3 = 6750$
Total	$81{,}600$
Demand factor from 220.32	
3 units = 45%	$\times 0.45$
	$36{,}720$
	$\div 240$
	$= 153 \text{ amperes}$

1 range, 8 kW, 240 volts

1 clothes dryer, 5000 watts, 240 volts

1 dishwasher, 600 watts, 120 volts

1 garbage disposal, 750 watts, 120 volts

1 water heater, 3000 watts, 240 volts

3 space heaters, 750 watts each 240 volts

The calculations are shown in Tables 6-6 and 6-7.

The calculation for two units using the standard method is larger than for three identical units using the optional method. Since it is permissible to use the smaller of the two calculations, 153 amperes per phase can be used. The neutral must be calculated by the standard method and would be the same in either case.

CHAPTER 6 TEST

1. A 2400 square foot single-family dwelling has how many 15 amp circuits for general lighting and receptacles?

2. What is the neutral load for a household electric dryer rated at 5500 watts?

3. How many appliance circuits are required for a single-family dwelling?

4. A 15 kW range would be calculated at how many watts for a service?

5. What is the wattage for the neutral in Problem 4?

6. What is the demand factor for four or more appliances fastened in place, other than electric ranges, clothes dryers, space-heating equipment, or air conditioning?

7. A 20 ampere laundry circuit is required for a single-family dwelling. What is the rating of this circuit in watts?

8. A 20 ampere circuit is required in the bathroom with no other outlets. What is the wattage of this circuit?

9. What is the most common noncoincidental load in a single-family dwelling?

10. In the optional calculation, how much of the general load, except for air conditioning and heating, is computed at 100%.

11. In a remodelling that does not add air conditioning or heat, how are the existing load and new load calculated?

12. What is the breakdown of the loads?

13. Using the optional method, calculate the phases for the following single-family load. (800 square feet):

1 dishwasher, 120 volt, 600 watts

1 oven, 12 kW

1 garbage disposal, 120 volt, 500 watts

1 water heater, 240 volts, 3 kW

1 attic fan, 250 watts, 120 volts

10 kW heat pump

1 5 kW dryer

14. What is the neutral size for Problem 13 in amperes?

15. What is the minimum size service for a single-family dwelling?

16. What is the service size in copper for Problems 14 and 15?

17. Calculate the total watts for the following single-family dwelling (1500 square feet):

12 kW oven

1 dishwasher, 120 volt, 600 watts

1 garbage disposal, 120 volt, 700 watts

1 water heater, 240 volts, 3 kW

1 clothes dryer, 4000 watts

1 heat pump, 10 kW

1 attic fan, 250 watts, 120 volts.

Use standard calculations.

18. What is the neutral size in watts for Problem 17?

19. What size aluminum service is needed for Problem 17?

20. In an optional calculation, all appliances are computed at nameplate rating with no deration factors applied. True or False?

chapter 7

MULTIFAMILY LOAD CALCULATIONS

7-1   STANDARD CALCULATIONS

A multifamily dwelling is a building or structure containing three or more dwelling units. The standard calculations for a multifamily dwelling are similar to the calculations for a single-family dwelling.

Included in the calculations are:

- 3 watts per square foot of the outside dimensions of the dwelling
- 2 small appliance circuits 1500 watts each
- 1 laundry circuit, 1500 watts each. Note that this circuit can be omitted if common laundry facilities are provided and available to all, or laundry facilities are not permitted to be installed in the units [see Table 210.52F, ex 1 and 2 of the *National Electrical Code®* (NEC)]. *
- Oven or cooktop
- Fixed appliances, dishwashers, garbage disposals, water heaters, and so on
- Heating or air conditioning, whichever is larger

The service to a single dwelling, including a single dwelling in a multifamily dwelling is no less than 100 amperes (Table 220.82A).

*This chapter contains material reprinted with permission from NFPA 70-2005, *National Electrical Code®*, Copyright © 2004, National Fire Protection Association, Quincy, MA 02269. This reprinted material is not the complete and official position of the NFPA on the referenced subject, which is represented only by the standard in its entirety. National Electrical Code and NEC are registered trademarks of the National Fire Protection Association, Quincy, MA.

Example. A multifamily building has eight units. Each unit has the following (laundry is not allowed in the unit):

840 square feet

1 oven, 12 kW

1 water heater, 2.5 kW, 240 volts

1 dishwasher, 840 watts, 120 volts

1 garbage disposal, 600 watts, 120 volts

1 central heat at 4 kW, 240 volts

See Table 7-1 for the calculations. 19,222 divided by 240 volts = 80 amperes on A phase, 18,982 divided by 240 volts = 79 amperes on B phase, and 10,922 divided by 240 volts = 46 amperes on the neutral. The service for this dwelling would be #4 copper conductors. Since the minimum service is 100 amperes, the neutral calculates to a #8 copper conductor; however, the authority having jurisdiction and/or the power company may require a minimum #6 copper neutral.

Table 7-1   Standard method, one unit

			Phase A	Phase B	Neutral
Lighting: 3 watts per sq ft	$840 \times 3 = 2520$				
Appliance circuits	$1500 \times 2 = 3000$				
Laundry circuit (omit)					
Subtotal		5520			
First 3000 watts at 100%		-3000	3000	3000	3000
Remainder at 35% up to		2520×0.35	882	882	882
120,000 watts					
Oven[a]	12,000		8000	8000	
70% of computed load for	8000	$\times 0.7$			5600
neutral					
Heat, 240 volt	4000		4000	4000	
1 Dishwasher, 120 volts[b]	840	$=$	840		840
1 Garbage disposal, 120 volts[b]	600	$=$		600	600
1 water heater, 240 volts[b]	2500	$=$	2500	2500	
Totals:			19,222	18,982	10,922
			$\div 240$	$\div 240$	$\div 240$
			$= 80$	$= 79$	$= 46$

[a]For the oven, NEC Table 220.55, column C applies.
[b]There are three fixed appliances in each unit, so Article 220.53 cannot be used.

Example. Eight units are supplied by a single feeder. Size the feeder supplying these dwelling units using the standard calculation. Laundry is not allowed in the units. Each unit has the following:

840 square feet

1 oven, 12 kW

water heater, 2.5 kW, 240 volts

1 dishwasher, 840 watts, 120 volts

1 garbage disposal, 600 watts, 120 volts

1 central heat at 4 kW, 240 volts

See Table 7-2 for the calculations.

Phase A = 92,206 ÷ 240 = 384 amperes

Phase B = 91,006 ÷ 240 = 379 amperes

Neutral = 42,146 ÷ 240 = 175 amperes

Table 7-2   Standard method, eight units

			Phase A	Phase B	Neutral
Lighting: 3 watts per sq ft	840 × 3 × 8 = 20,160				
Appliance circuits	1500 × 2 × 8 = 24,000				
Laundry circuit (omit)					
Subtotal	44,160				
First 3000 watts at 100%	− 3000		3000	3000	3000
Remainder at 35% up to 120,000 watts	41,160 × 0.35		14,406	14,406	14,406
Oven[a]	23,000			23,000	23,000
70% of computed load for neutral	23,000	× 0.7			16100
Heat, 240 volt	4000	× 8		32,000	32,000	
1 Dishwasher, 120 volts[b]	840	× 8 =	× 0.75	5040		5040
1 Garbage disposal, 120 volts[b]	600	× 8 =	× 0.75		3600	3600
1 Water heater, 240 volts[b]	2500	× 8 =	× 0.75	15,000	15,000	
Totals:				92,446	91,006	42,146
				÷ 240	÷ 240	÷ 240
				= 384	= 379	= 176

[a]For the oven, NEC, Table 220.55, column C applies (eight ovens, 12 kW or less).
[b]There are three fixed appliances in each unit; however, when combined, there are 24 fixed appliances and Table 220.53 applies.

Wire size is obtained from Table 310.16, THWN conductors. Parallel 3/0 copper conductors in separate conduits would be sufficient with a 1/0 neutral in each conduit, since 1/0 is the minimum size that can be run in parallel. This would bring 400 amperes to the building. Parallel 250 mcm copper conductors in a single conduit would also work well. Each 250 mcm copper conductor is rated at 255 amperes, which added together equals 510 amperes. Since there will be four current-carrying conductors in the same conduit, the conductors must be derated to 80%. This yields 408 amperes (Table 310.15B2a). Since the neutral is in a single conduit, a single conductor can be used that will carry the 175 amperes required. A 2/0 THWN copper conductor will be sufficient for the neutral. Since the neutral carries only the unbalanced current from the other conductors, it does not need to be derated as a current-carrying conductor (Table 310.15B4).

Example. This example is a service to a 32 unit multifamily dwelling with the following (laundry is not allowed in the unit):

840 square feet

1 oven, 12 kW

1 water heater, 2.5 kW, 240 volts

1 dishwasher, 840 watts, 120 volts

1 garbage disposal, 600 watts, 120 volts

1 central heat at 4 kW, 240 volts

See Table 7-3 for the calculations. Phase A + Phase B = 313,270 ÷ 240 = 1305 amperes. Phase B = 307,510 watts/240 volts = 1281 amperes. Since the neutral is only carrying the imbalance of the phase to the neutral load, and is a part of a two-phase, three-wire system, a further demand factor of 70% can be applied to that portion of the neutral that exceeds 200 amperes:

$$
\begin{array}{rl}
12{,}570 \div 240 = & 523 \text{ amperes} \\
-200 \times 100\% = & 200 \\
323 \times 70\% = & \underline{226} \\
& 376 \text{ amperes}
\end{array}
$$

The wire sizes can be configured in several ways in this example. In the preceding example, all of the units were identical.

Table 7-3   Standard method, 32 units

			Phase A	Phase B	Neutral
Lighting: 3 watts per sq ft	$840 \times 3 \times 32 = 80,640$				
Appliance circuits	$1500 \times 2 \times 32 = 96,000$				
Laundry circuit (omit)					
Subtotal		176,640			
First 3000 watts at 100%		$-3000 \times 100\%$	3000	3000	3000
Remainder at 35% up to		173,640			
120,000 watts		$-117,000 \times 35\%$	40,950	40,950	40,950
Remainder at 25%		$56,640 \times 25\%$	14,160	14,160	14,160
Oven, 12 kW[a]	47,000			47,000	47,000
Neutral for cooktop, 70%	47,000	$\times 70\%$			32,900
1 Dishwasher, 120 volts[b]	840	$\times 32 =$	$\times 70\%$	20,160	20,160
1 Garbage disposal, 120 volts[b]	600	$\times 32 =$	$\times 75\%$		14,400	14,400
1 Water heater, 240 volts[b]	2500	$\times 32 =$	$\times 75\%$	60,000	60,000
Heat, 4 kW, 240 volts	4000	$\times 32 =$		128,000	128,000
Totals:			313,270	307,510	125,570
			$\div 240$	$\div 240$	$\div 240$
			$= 1305$	$= 1281$	$= 523$

[a]Oven is calculated from column C of Table 220.55—32 units = 15 kW + 1 kW for each unit (32).
[b]There are three fixed appliances in each unit; however, when combined, there are 96 fixed appliances and Article 220.53 applies.

7-2   OVENS OF UNEQUAL RATING

A multifamily dwelling can have different floor plans and different appliances. If there are different-size ranges involved and the rating is between 12 kW and 27 kW, the ranges can be added together and then divided by the number of ranges to get an average. If the 32 units in the preceding problem had the following ranges:

$$8 \text{ at } 12 \text{ kW} = 96 \text{ kW}$$
$$8 \text{ at } 15 \text{ kW} = 120 \text{ kW}$$
$$8 \text{ at } 18 \text{ kW} = 144 \text{ kW}$$
$$\underline{8 \text{ at } 20 \text{ kW} = 160 \text{ kW}}$$
$$520 \div 32 = 16.25 \text{ kW average}$$

32 ranges at 16 kW would be 15 kW + 32 kW = 47 kW. Since 16 kW exceeds 12 kW by 4 kW, there is 5% of 47 kW for each 1 kW in excess. $4 \times 5 = 20$. 47 kW $\times 1.20 = 56,400$ watts.

7-3   HOUSE LOADS FOR MULTIFAMILY DWELLINGS

House loads in a multifamily dwelling are loads that are common to all tenants. These loads can consist of swimming pools, exterior lighting, office area, club house, and laundry facilities to name a few. This portion of the load is computed separately from the dwelling units themselves. The following example is a house load consisting of the following:

15 by 20 square foot room

12 120 volt washing machines

8 5000 watt, 240 volt clothes dryers

This load must be calculated separately but is added to the service feeding the multifamily dwelling in either the standard or optional calculation. A house load has its own meter and is paid by management. Table 220.12 does not spell out this type of occupancy, so 2 watts per square foot was selected for minimum lighting. It would be prudent to get a ruling from the authority having jurisdiction to find out what is allowed in this type of dwelling. The continuous load rule would apply to the lighting since it would be on 3 hours or more.

15 by 20 = $300 \times 2 \times 1.25 = 750$ watts

12 12 ampere washing machines, 120 volts, $12 \times 12 \times 120 = 17,280$ watts

Largest motor $12 \times 120 = 1440 \times 0.25 = 360$ watts

8 5000 watt, 240 volts clothes dryers = 40,000 watts

Neutral load for dryers, $40,000 \times 0.7 = 28,000$ watts

See Table 7-4 (next page) for the calculations. The phases would increase the service by 243 amperes and the neutral would increase it by 193 amperes. If the increased neutral size is above 200 amperes total, a further reduction of 70% would be allowed in excess of 200 amperes since all of the neutral is carrying the unbalanced load.

7-4   OPTIONAL CALCULATION FOR MULITIFAMILY DWELLING

The optional calculation method must meet the following conditions:

Table 7-4   House load

		Phases	Neutral
Lighting: 2 watts per sq ft	$300 \times 2 \times 1.25 =$	750	570
Washing machines, 12 amps, 120 volts	$12 \times 12 \times 120 =$	17,240	17,240
Largest motor = 360 watts		360	360
Clothes dryers, 5000 watts, 240 volts, 8	$5000 \times 8 =$	40,000	
Neutral at 70%	$40,000 \times 0.7 =$		28,000
		58,350	46,350
		÷ 240	÷ 240
		= 243	= 193

- Only one feeder per dwelling
- Electric cooking equipment must be in each dwelling. However, if electric cooking equipment is not in each unit, the standard calculation without electric cooking equipment can be compared to the optional calculation method with the same load plus electric cooking equipment by adding 8 kW per unit, the smaller of the two calculations can be used.
- Electric space heating, air conditioning, or both is included in each unit.

The connected load to which the demand factor of Table 220.84 applies includes the following.

- Lighting and general use receptacles is calculated at 3 watts per square foot
- Two small appliance circuits of 1500 watts each
- 1500 watts for each laundry branch circuit. Note: The laundry circuit may be omitted under the provisions of Article 210-52e, exception 1 or 2.
- The nameplate rating of all fixed appliances that are permanently connected, including cooking equipment, (ranges, wall-mounted ovens, counter-mounted cooktop units, clothes dryers, and space heaters.
- The nameplate rating of all motors and of all low-power factor loads.
- The larger of the air conditioning and the space heating loads.

Example. A multifamily building has eight units. When calculating the service or feeder to one unit, the optional method for a single family is used (220). Each unit has the following:

840 square feet

1 oven at 12 kW

1 water heater at 240 volts, 2.5 kW

1 120 volt dishwasher at 840 watts

1 garbage disposal at 120 volts, 600 watts

1 heater at 4 kW, 240 volts

See Table 7-5 for the calculations. Each unit was calculated as a single unit. The total is 17,184 watts. Divide 17,184 by 240 = 72 amperes. Since the unit is a single-family unit, all the rules for a single-family dwelling apply. Article 220-82 states that the feeder must be 100 amperes or greater.

The neutral must be calculated by Article 220-61, which is the standard method.

Make a list of the line to neutral loads in the above example:

$$
\begin{array}{ll}
\text{Square footage} & 840 \times 3 = 5520 \\
\text{2 appliance circuits} & \underline{-3000} \times 100\% = 3000 \\
& 2520 \times 35\% \;\; = \underline{\;\; 882} \\
& 3882
\end{array}
$$

The oven is computed from Table 220.55, Column C, 12 kW range = 8 kW. Remember that the neutral will be 70% of the ungrounded conductors:

Table 7-5 Optional calculation, multifamily dwelling, 1 unit

Lighting, 3 watts per sq ft	$840 \times 3 \times 1$	2520
Appliance circuits	$1500 \times 2 \times 1$	3000
Laundry circuit (omit)		
Oven 12 kW	$12,000 \times 1 \times 1$	12,000
Water heater, 2.5 kW, 240 volts	$2500 \times 1 \times 1$	2500
Dishwasher, 840 watts	$840 \times 1 \times 1$	840
Garbage disposal, 500 watts, 120 volts	$600 \times 1 \times 1$	600
		21,460
		$-10,000 \times 100\% =$ 10,000
		$11,460 \times 40\% \;\;\; =$ 4584
		14,584
Heat, 4 kW, 240 volts	$400 \times 1 \times 0.65$	2600 2600
Total		17,184

8000 watts × 70% = 5600 watts
Dishwasher = 840 watts
Garbage disposal = 600 watts

Bring all of the 120 volt loads together and add them up:

Lighting	3882
Oven	5600
Dishwasher	840
Garbage disposal	600
Total neutral	10,922 watts

10922/240 = 46 amperes

Now size the service to the building containing the eight units. Make a list of all the loads.

Example. Optional calculation, eight units. From Table 220.84, go down to 8-10 in number of dwelling units. In the column directly across find the demand factor percentage. Take the percentage times the total connected load just calculated above:

$$
\begin{array}{r}
203,680 \\
\times\, 0.43 \\
\hline
87,582
\end{array}
$$

Next, divide the new watts (87,582) by 240 volts to yield the amperage in the ungrounded conductors:

87582/240 = 365 amperes

Table 7-6   Optional calculation, multifamily dwelling, eight units

Lighting, 3 watts per sq ft	840 × 3 × 8	20,160
Appliance circuits	1500 × 2 × 8	24,000
Laundry circuit (omit)		
Oven, 12 kW	12,000 × 1 × 8	96,000
Water heater, 2.5 kW, 240 volts	2500 × 1 × 8	20,000
Dishwasher, 840 watts	840 × 1 × 8	6720
Garbage disposal, 500 watts, 120 volts	600 × 1 × 8	4800
Heat, 4 kW, 240 volts	4000 × 1 × 8	32,000
Total		203,680

Make a list of the line-to-neutral loads.

Lighting
Appliance circuits
Oven
Dishwasher
Garbage disposal

First, compute the lighting and small appliance circuits:

$$840 \times 3 \times 8 = 20{,}160$$
$$2 \times 1500 \times 8 = \underline{24{,}000}$$
$$44{,}160$$

Use Table 220.42. The first 3000 watts are at 100%; the remainder (up to 117,000 watts) are at 35%:

$$44{,}160$$
$$\underline{- 3000} \times 100\% = \quad 3000$$
$$41{,}160 \times 0.35\% = \underline{14{,}406}$$
$$17{,}406 \text{ watts}$$

The ovens are computed by going to Table 220.55. Look down the column number of appliances. Go to 8 in the first column (number of appliances), then go across to column C. Maximum demand for ranges not over 12 kW is 23 kW for 8 ranges.

This result is the value of the ungrounded conductors. To find the neutral, multiply the result by 70%:

$$23{,}000 \times 70\% = 16{,}100$$

The next load is the garbage disposal and dishwasher. Combine the two for a total of 1440 watts:

$$1440 \times 8 \times 0.75 = 8640 \text{ watts}$$

Article 220-53 states that for four or more fixed appliances in a one-family, two-family, or multifamily dwelling, a demand factor of 75% may be applied if they are only on one feeder. This does not include electric ranges, clothes dryers, space-heating equipment, or air conditioning. This demand factor of 75% can only be applied during a standard calculation and only affects the neutral since the ungrounded conductors have previously been calculated by the optional method. For a neutral load,

$$
\begin{aligned}
\text{Lighting load} &= 17,406 \\
\text{Oven} &= 16,100 \\
\text{Fixed appliances} &= \underline{8640} \\
& 42,146 \text{ watts/240 volts} = 176 \text{ amperes}
\end{aligned}
$$

Example. Optional calculation, 32 units. A service is needed to a 32 unit building. Calculate the service size in amperage for the phases and neutral. Each unit has the following:

840 square feet

1 oven at 12 kW

1 water heater at 240 volts, 2.5 kW

1 120 volt dishwasher at 840 watts

1 garbage disposal at 120 volts, 600 watts

1 heater at 4 kW, 240 volts

The calculations are in Table 7-7. The total is 814,720 watts. Table 220.84 shows the demand factor for 32 units to be 31%:

$$814,720 \times 0.31 = 252563 \text{ watts} \div 240 = 1052 \text{ amperes}$$

Table 7-7 Optional calculation, multifamily dwelling, 32 units

Lighting, 3 watts per sq ft	$840 \times 3 \times 32 =$ 80,640
Appliance circuits	$1500 \times 2 \times 32 =$ 96,000
Laundry circuit (omit)	
Oven, 12 kW	$12,000 \times 1 \times 32 =$ 384,000
Water heater, 2.5 kW, 240 volts	$2500 \times 1 \times 32 =$ 80,000
Dishwasher, 840 watts, 120 volts	$840 \times 1 \times 32 =$ 26,880
Garbage disposal, 500 watts, 120 volts	$600 \times 1 \times 32 =$ 19,200
Heat, 4 kW, 240 volts	$4000 \times 1 \times 32 = \underline{128,000}$
Total	814,720

To size the neutral, list all of the neutral loads:

Lighting
Appliance circuits
Oven
Dishwasher
Garbage disposal

First compute the lighting and small appliance circuits:

$$840 \times 3 \times 32 = 80,640$$
$$2 \times 1500 \times 32 = \underline{96,000}$$
$$176,640$$

Use Table 220.42. The first 3000 watts are at 100%; the next 117,000 watts are at 35%; the remainder above the first 120,000 watts (3000 + 117,000) is at 25%:

$$176,640$$
$$\underline{-3000} \times 100\% = \quad 3000$$
$$173,640$$
$$\underline{-117,000} \times 35\% = 40,950$$
$$56,640 \times 25\% = \underline{14,160}$$
$$58,110$$

The ovens are computed by going to Table 220.55. Look down the column number of appliances. Go to 32 in the first column (number of appliances), then go across to column C. Maximum demand for ranges not over 12 kW is 15 kW + 1 kW for each range (32) = 47 kW.

This result is the value of the ungrounded conductors. To find the neutral, multiply the result times 70%:

$$47,000 \times 70\% = 32,900$$

The next load is the garbage disposal and dishwasher:

$$840 \times 32 = 26,880 \text{ watts}$$

$$600 \times 32 = 19,200 \text{ watts}$$

Article 220.53 states that for four or more fixed appliances in a dwelling unit on one feeder, whether for a one-family, two-family,

Table 7-8 Neutral calculation, multifamily dwelling, 32 units

Lighting, 3 watts per sq ft	$840 \times 3 \times 32$	80,640	
Appliance circuits	$1500 \times 2 \times 32$	96,000	
Laundry circuit (omit)			
		176,640	
		$-3000 \times 100\%$ =	3000
		173,640	
		$-117,000 \times 35\%$ =	40,950
		$56,640 \times 25\%$	14,160
Oven, 12 kW	$47,000 \times 0.7$		32,900
Dishwasher, 840 watts	$840 \times 1 \times 32$	$26,880 \times 75\%$ =	20,160
Garbage disposal, 600 watts	$600 \times 1 \times 32$	$19,200 \times 75\%$ =	14,400
Total			125,570

or multifamily dwelling a demand factor of 75% may be applied. This does not include electric ranges, clothes dryers, space-heating equipment, or air conditioning. This demand factor of 75% can only be applied during a standard calculation and only affects the neutral since the ungrounded conductors have previously been calculated by the optional method:

$$26,880 \times 0.75 = 20,160$$

$$19,200 \times 0.75 = 14,400$$

See Table 7-8 for the calculations. The total neutral load calculates to 125,570 watts ÷ 240 = 523 amperes. Since this neutral carries only the imbalance between phases, a further reduction is allowed (Table 220.61). The first 200 amperes is calculated at 100% and the remainder at 70%:

$$
\begin{array}{l}
523 \\
-200 \times 100\% = 200 \\
\underline{323 \times 70\% = 226} \\
446 \text{ amperes}
\end{array}
$$

7-5 LAUNDRY IN EACH UNIT

If laundry is allowed in each dwelling, the house load for laundry would not be used. Other house loads could include swimming pools, outside lighting, clubhouses, and storage areas, to name a few.

If washers and clothes dryers were permitted in each of the previous examples, they would add substantially to the size of each service.

In a one-unit standard calculation (see Example, page 128), the ungrounded conductors were 85 amperes and the neutral was 46 amperes. To add a washer and dryer to the unit, the laundry circuit is 1500 watts. All of the deration has been done, so 1500 watts would be multiplied by 35%:

$$1500 \times 0.35 = 525$$

The clothes dryer is 5000 watts and would be 100% on the phases and 70% on the neutral.

Phases	Neutral
5000	$5000 \times 0.70 = 3500$
+ 525	+ 525
$5525 \div 240 = 23$ amperes	$4025 \div 240 = 17$ amperes

The phases would increase by 23 amperes to 103 amperes and the neutral would increase by 17 amperes to 63 amperes.

In the Example on page 129, the eight-unit standard calculation, eight 1500 watt circuits would be needed:

$$8 \times 1500 = 12,000 \text{ watts}$$

Since the original lighting and receptacle load was well below 120,000 watts, the demand factor of 35% would apply. Electric clothes dryers are 5000 watts each. Table 220.54 allows eight household electric clothes dryers to be calculated at 60% of the dryers' value when one service or feeder is involved:

$$1500 \times 8 \times 0.35 = \quad 4200$$
$$5000 \times 8 \times 0.60 = \underline{24,000}$$
$$28,200 \div 240 = 118 \text{ amperes}$$

The neutral for the clothes dryers would be 70% of the connected load, which would be $24,000 \times 0.70 = 16,800$ watts plus the laundry circuit of 4200 watts:

$$4200$$
$$\underline{16,800}$$
$$21,000 \text{ watts} \div 240 = 88 \text{ amperes}$$

This would increase the load on the ungrounded conductors from 399 amperes to 517 amperes.

The neutral would increase from 88 amperes to 263 amperes. The neutral would have a further reduction of 70% for the amount above 200 amperes:

$$
\begin{array}{r}
263 \\
-\ 200 \times 100\% = 200 \\
\hline
63 \times 70\% \ = \ \underline{44} \\
244 \text{ amperes}
\end{array}
$$

In the Example on page 130, the amperage for the ungrounded conductors was calculated at 1361 amperes and the neutral was calculated at 423 amperes. To add a washer and dryer to each of these units, 32 – 1500 watt laundry circuits must be added:

$$1500 \times 32 = 48,000 \text{ watts}$$

Since the original calculation was above 120,000 watts for the general lighting and receptacle load, a demand factor of 25% can be used:

$$48,000 \times 0.25 = 12,000$$

A clothes dryer is 5000 watts. Table 220.54 allows 32 clothes dryers to be derated by the following formula:

$$35\% - [0.5 \times (\text{number of dryers} - 23)] = 35\% - [0.5 \times (32 - 23)]$$
$$= 35 - (0.5 \times 9) = 35\% - 4.5 = 30.5\%$$

$$5000 \times 32 \times 0.305 = 48800 \text{ watts}$$

$$\text{Neutral} = 48,800 \times 70\% = 34,160 \text{ watts}$$

$$
\begin{array}{ll}
\text{Laundry circuit} & 12,000 \\
\text{Dryer} & \underline{48,800} \\
& 60,800 \div 240 = 253 \text{ amperes per phase}
\end{array}
$$

$$
\begin{array}{l}
\text{Neutral } 48,800 \times 70\% = 34,160 \\
\phantom{\text{Neutral } 48,800 \times 70\% =}\ \underline{12,000} \\
\phantom{\text{Neutral }}46,160 \text{ watts} \div 240 = 192 \text{ amperes}
\end{array}
$$

The ungrounded conductors would be increased by 253 amperes. Because the neutral has already been derated above 200 amperes, the new neutral load can be derated by 70% as well: 192 × 70% = 134 amperes.

In the Example on page 134, optional calculation for a single-family dwelling, the appliance circuit must be added in as well as the dryer at 100%. Since the load has already been calculated, a derating of 40% of the added load will be sufficient:

$$1500 + 5000 \times 40\% = 2600 \div 240 = 11 \text{ amperes}$$

The neutral would be sized the same as in the Example on page 130 (standard method), which yielded 17 amperes.

In the Example on page 135, optional calculation for eight multifamily units, the ungrounded conductors were calculated at 359 amperes. In an optional calculation, the laundry circuit and the clothes dryer are added into the calculation at 100% of their value. Then, a demand factor (Table 220.84) of 43% is applied:

$$1500 \times 8 = 12,000 \text{ watts}$$
$$5000 \times 8 = \underline{40,000} \text{ watts}$$
$$52,000 \text{ watts} \times 43\% = 22360 \text{ watts} \div 240 = 93 \text{ amperes}$$

The neutral would be calculated the same way as the neutral for the eight unit-standard calculation method, which was 244 amperes.

In the Example on page 137, optional calculation for 32 units, the ungrounded conductors were calculated at 1051 amperes. In an optional calculation, the laundry circuit and clothes dryers are added in at 100%, then a demand factor (Table 220.84) of 31% would is applied:

$$1500 \times 32 = 48,000 \text{ watts}$$
$$5000 \times 32 = \underline{160,000} \text{ watts}$$
$$208,000 \text{ watts} \times 31\% = 64480 \div 240 = 269 \text{ amperes}$$

This would be added to the 1051 amperes previously calculated: 1051 + 269 = 1320 amperes.

The neutral is calculated by the standard method (see Example on page 130) for 32 units and adding 134 amperes, bringing the neutral up to 557 amperes.

7-6 OPTIONAL CALCULATION WITHOUT ELECTRIC COOKING

In the optional calculation without electric cooking (gas cooking is used instead), Article 220.84 allows the service to be calculated by the standard method without electric cooking and the optional method with electric cooking, based on 8 kW per unit and using the smaller of the two loads.

Example. A 32-unit dwelling has the following:

840 square feet per unit

1 dishwasher, 120 volts, 840 watts

1 garbage disposal, 120 volts, 600 watts

1 water heater, 240 volts, 2500 watts

Electric heat, 240 volts, 4000 watts

Laundry in units is not allowed and there is no electric cooking.

See Table 7-9 for the calculations. The phases are 1098 amperes. The neutral is 314 amperes. However, the neutral can be further reduced because it exceeds 200 amperes:

Table 7-9 Standard method, 32 units without electric cooking

				Phase A	Phase B	Neutral
Lighting: 3 watts per sq ft	$840 \times 3 \times 32 =$		80,640			
Appliance circuits	$1500 \times 2 \times 32 =$		96,000			
Laundry circuit (omit)						
Subtotal			176,640			
First 3000 watts at 100%			− 3000	3000	3000	3000
Remainder at 36% up to			173,640			
120,000 watts			− 117,000	40,950	40,950	40,950
Remainder at 25%			$56,6,40 \times 0.25$	14,160	14,160	14,160
Heat, 240 volt	4000	× 32		128,000	128,000	
1 Dishwasher, 120 volts[a]	840	× 32 =	× 0.75	20,160		20,160
1 Garbage disposal, 120 volts[a]	600	× 32 =	× 0.75		14,400	14,400
1 Water heater, 240 volts[a]	2500	× 32 =	× 0.75	60,000	60,000	
Totals:				266,270	260,510	92,670
				÷ 240	÷ 240	÷ 240
				= 1109	= 1085	= 386

[a]There are three fixed appliances in each unit; however, when combined, there are 96 fixed appliances and Article 220.53 applies.

Table 7-10 Optional calculation, multifamily dwelling, 32 units with 8 kW cooktops

Lighting, 3 watts per sq ft	840 × 3 × 32	80,640
Appliance circuits	1500 × 2 × 32	96,000
Laundry circuit (omit)		
Oven, 8 kW	8000 × 1 × 32 =	256,000
Water heater, 2.5 kW, 240 volts	2500 × 1 × 32 =	80,000
Dishwasher, 840 watts	840 × 1 × 32 =	26,880
Garbage disposal, 500 watts, 120 volts	600 × 1 × 32 =	19,200
Heat, 4 kW, 240 volts	4000 × 1 × 32 =	128,000
Total		= 686,720

$$\begin{array}{r} 386 \text{ amperes} \\ -\underline{200 \times 100\% = 200} \\ 186 \times 70\% \quad = \underline{130} \\ 330 \text{ amperes} \end{array}$$

See Table 7-10 for the calculations. The total of 686,720 watts is multiplied by the demand factor for 32 units (Table 220.84) of 31%:

$$686{,}720 \times 31\% = 212{,}883 \div 240 = 887 \text{ amperes}$$

The optional calculation with electric cooking is less (887 amperes) than the standard calculation without electric cooking (1109 amperes). The optional calculation can be used. The neutral would be the same as the standard calculation in Table 7-10 of 330 amperes.

CHAPTER 7 TEST

1. What is the minimum size service for a single unit in a multifamily dwelling?

2. Is a laundry circuit required in all multifamily dwellings? Why?

3. An eight-unit multifamily dwelling has one service, the cooking equipment in each unit is rated at 7 kW. What is the minimum load in the standard calculation for the cooking units in the service in watts?

4. What is the minimum load in Problem 3 for the neutral?

5. In Problem 3, what would be the minimum wattage for one oven in a service feeding one unit?

6. What would be the calculated neutral load for Problem 5 in watts?

7. A 16-unit multifamily dwelling has an allowance of 5 kW per unit for electric clothes dryers. What would be the minimum calculated load for 16 clothes dryers on one service in watts?

8. What is the calculated load on the neutral for Problem 7 in watts?

9. Each unit of an eight-unit apartment building has the following: 600 square feet; 1 garbage disposal, 120 volt, 750 watts; 1 dishwasher, 600 watts, 120 volts; 1 water heater, 240 volts, 2500 watts; 1 cooking unit, 8 kW. Laundry is not allowed in the units. What is the size of the service to each unit by the standard calculation? Give your answer in amperes.

10. What is the size of the service for the eight units? Use the standard calculation and answer in amperes.

11. Each unit of a 32-unit multifamily dwelling has the following: 700 square feet; 1 cooking unit, 8 kW; 1 garbage disposal, 120 volts, 500 watts; 1 dishwasher, 120 volts, 600 watts; 1 water heater, 240 volts, 3000 watts; electric heat, 2250 watts, 120 volts. What are the phases for the 32 units using the optional calculation? Laundry is not allowed in each unit.

12. What is the minimum size of the neutral in Problem 11?

13. In a multifamily dwelling, what is the house load?

14. Can house loads be added to multifamily dwelling loads and be derated by the optional calculation?

15. A multifamily dwelling has 20 units. If each unit is 450 square feet, what is the lighting load including the appliance circuits? Laundry is not allowed in units. Use the standard calculation.

16. What is the lighting load, including appliance circuits, in Problem 15 using the optional method?

17. Can the optional calculation be used if the dwellings do not have cooking equipment?

18. A multifamily dwelling has three sizes of units. Unit A has a 12 kW electric range, unit B has a 14 kW electric range, and unit C has a 15 kW range. If there are three unit A's, four unit B's, and three unit C's, what is the average range size in order to compute the load?

19. What is the neutral portion of the load for the average range in Problem 18?

20. A house load has outside lighting of 30 poles, each with a 150 watt incandescent lamp. What is the amperage computed with the multifamily service?

21. What is the initial load for 5 kW heating using the optional calculation for four units?

22. What is the load for Problem 21 after deration using the optional calculation?

23. How many 15 ampere circuits for lighting and receptacles are there in a multifamily dwelling of 900 square feet?

24. What is the demand load for 24 cooking units using the standard calculation if each unit is 12 kW?

25. What is the neutral load for Problem 24?

chapter 8

COMMERCIAL CALCULATIONS

8-1   GENERAL

Commercial calculations include a wide range of applications, including schools, hospitals, manufacturing facilities, office buildings, gas stations, hotels, and shopping centers, to name a few. It would not be uncommon to have different types of occupants under the same roof. For instance, a large hotel may have a beauty shop, restaurant, conference rooms, stores, and offices as well as guest rooms, halls, corridors, closets, stairways, and storage spaces. There are several different commercial applications, differing in size and requirements. Some basic rules to remember in commercial wiring are well worth noting and reviewing:

National Electrical Code (NEC), Article 210.20A, Continuous Duty Rule*

Article 220.12 and Table 220.12, Lighting loads for specified occupancies

Table 220.44, Receptacle loads for occupancies other than dwellings

Article 220.50, Motors, 125% of largest

Article 220.60, Noncoincidental loads

Article 220.61, Neutral loads

Article 310-4, Parallel conductors

*This chapter contains material reprinted with permission from NFPA 70-2005, *National Electrical Code®*, Copyright © 2004, National Fire Protection Association, Quincy, MA 02269. This reprinted material is not the complete and official position of the NFPA on the referenced subject, which is represented only by the standard in its entirety. National Electrical Code and NEC are registered trademarks of the National Fire Protection Association, Quincy, MA.

Other rules governing conduit size, special conditions, wire size, and other factors still apply.

The continuous load rule, Article 210.20A, especially applies to lighting, motors, air conditioners, heating, and water heaters. Remember, a continuous load is any load that is likely to be on for 3 hours or more.

General lighting loads by occupancy, Table 220.12, are self-explanatory. The square footage of an occupancy is multiplied by the unit load, which yields watts per square foot. If the lighting is already designed, then compare the watts per square foot with the design value and use the larger of the two for calculations. The minimum lighting allowed is the general load by occupancy. However, if the design lighting is larger, then it must be used in the calculation.

Also bear in mind that a building may have more than one type of occupancy. An office/warehouse is a good example. The space used for the office would be 3.5 watts times the area of the office. The warehouse area is multiplied by ¼ watt per square foot. The office lighting at 3.5 watts per square foot and the warehouse lighting at ¼ watt per square foot are added together and multiplied at 125% for continuous load and the result, if larger than the design value, is the total wattage needed for the service conductors.

Example: 1800 square foot building. No lighting on the building plan. 50% office, 50% warehouse:

$$900 \times 3.5 = 3150$$
$$900 \times ¼ = \underline{225}$$
$$3375 \times 1.25 = 4219 \text{ watts}$$

The minimum lighting requirements for this building would be 4219 watts. This does not mean that the lighting has to equal 4219 watts; however, the service has to allow 4219 watts for the lighting.

Article 220.44 shows how to calculate receptacle loads. There is no minimum requirement for duplex receptacles in commercial applications. Also, in dwellings, receptacles are included in the lighting demand factor; in commercial buildings, receptacles are calculated separately. In commercial applications, a receptacle load in excess of 10 kW can be derated in accordance with Table 220.44.

Article 220.50 concerns motors and refers to other articles pertinent to this area. These rules need to be applied as they arise.

Article 220.60 concerns noncoincident loads. The most common application of this rule is heating and air conditioning. Simply stated, whichever load is larger will be used in the calculation. Since air conditioners contain motors, 125% of the connected load is applied. Article 220.51 states that fixed electric space heating shall be computed at 100%. The smaller of the two loads may be dropped as both the air conditioning and the heat are unlikely to be on at the same time.

Article 220.61 concerns the calculation of the neutral. In larger installations, the neutral could easily exceed 200 amperes, of which the excess can be derated. However, care should be taken, especially in commercial buildings, because part of the neutral may be used for nonlinear loads (electronic equipment, electronic/electric-discharge lighting, adjustable-speed-drive systems, and similar equipment). The high percentage of lighting and electronics in commercial buildings will not allow the neutral to be derated, so even though the neutral may exceed 200 amperes, a reduction may not be possible.

Article 310-4 concerns parallel conductors. This is an important article. Many times, it is cheaper to run parallel conductors than bigger single conductors. Doubling the size of the conductor does not double the ampacity. For example:

2 500 Mcm Thw copper equals 760 amperes

1 1000 Mcm Thw copper equals 545 amperes

In this case, although the wire size is equal—1000 circular mills of copper—the amperage is not. The two parallel conductors have a 30% higher ampacity than the larger single conductor. That is because there is more surface area in the two 500 Mcm conductors. This makes parallel conductors very important in larger installations.

8-2 SINGLE-PHASE SMALL COMMERCIAL

To do the calculations for a building or part of a building, a format must be developed. After students are familiar with calculating, they may develop their own methods.

Problem. A 2000 square foot office/warehouse has 800 square ft of office space; the remainder is warehouse space. Lighting is eight 2 × 4 fluorescent lay-in fixtures in the office, 200 watts each. The warehouse lighting is four 8 foot strip lights, 150 watts each. The air conditioner is a 2 ton unit, drawing 24 amperes at 240 volts. Heating is equal to 10 kW, also at

240 volts. There are 10 convenience outlets in the office and four in the warehouse. The service for this occupancy is 120/240 volt, single phase. Calculate the load.

First, size the wire in THW copper conductors, phase and neutral.

Step 1: Lighting

$$
\begin{array}{ll}
\text{A.} & 800 \times 3.5 \times 125\% = 3500 \\
& 1200 \times \frac{1}{4} \times 125\% = \underline{375} \\
& \phantom{1200 \times \frac{1}{4} \times 125\% = } 3875
\end{array}
$$

$$
\begin{array}{ll}
\text{B.} & 8 \times 200 \times 125\% = 2000 \\
& 4 \times 150 \times 125\% = \underline{750} \\
& \phantom{4 \times 150 \times 125\% = } 2750
\end{array}
$$

For A, the minimum code value, 3875 watts, is larger than the design of 2750 watts.

Step 2: Convenience outlets

10 office
<u>4</u> warehouse
14 × 180 = 2520 watts

Step 3: Heating and air conditioning

Air conditioning—24 × 240 × 1.25 = 7200 watts

Heating—10,000 × 100% = 10,000 watts

Heating is larger, so air conditioning would be dropped. Now combine all the values that apply to the service. The load calculation is:

800 square foot office

1200 square foot warehouse

14 convenience outlets

10 kW heat

See Table 8-1 for the calculations. Divide the watts by 240 volts to yield the amperage:

$$16395 \div 240 = 68 \text{ amperes}$$

Table 8-1   Office/warehouse combination

		Phase	Neutral
Office lighting	$800 \times 3.5 \times 1.25 =$	3500	3500
Warehouse lighting	$1200 \times 0.25 \times 1.25 =$	375	375
Receptacles	$14 \times 180 =$	2520	2520
Heat	1000×1	10,000	
Phases		16,395	6395
	$16,395 = 68.3$		
	$\div 240$		
Neutral	$6395 = 26.6$		
	$\div 240$		

The ungrounded conductors of your service must be at least 68 amperes. To size the neutral, convert all of the phase-to-neutral loads to amperage, which in this case will be lighting and receptacles (from Table 8-1, 27 amperes). The nearest wire size for Phase A is #4, for Phase B it is #4, and for Neutral it is #6. The neutral is #6 because the local power company requires a minimum #6 for the neutral.

8-3   THREE-PHASE SMALL COMMERCIAL

Three-phase power is required in a larger installation. Minimum requirements may vary around the country, from power company to power company, but at least one 3.5 horsepower motor is required. There are two types of three-phase power available. The first is the 208/120 volt, three phase, four wire wye system. The second is the 240/208/120 volt high-leg delta system. Both these systems were discussed earlier.

With regard to load calculation, the 208/120 volt wye system can have lighting balanced on all three phases. It is fairly easy to balance the lighting since there is no difference between phase to ground at any point. The 240/208/120 volt high-leg delta system is a bit more complicated. If a three-phase panel is used, care must be taken to ensure that none of the 120 volt load ends up on the high leg. Only two phases to ground can be used for lighting and receptacle usage. The power company would elect to use the 240/208/120 volt grounded delta. If the three-phase load were small, they would probably use the open delta.

Calculate the load on the following:

1000 square foot office

3000 square foot manufacturing area

With the following equipment:

    1 5 horsepower, three-phase air compressor

    1 3 horsepower, three-phase conveyor

    1 10 horsepower, three-phase lathe

    1 3 horsepower, three-phase grinder

    1 1 horsepower, 110 volt drill press

    1 ½ horsepower, 110 volt blower

    1 ¾ horsepower, 110 volt blower

    10 110 volt receptacles in the office

    15 110 volt receptacles in the manufacturing area

    10 kW heat strips, three-phase

    2½ ton air conditioning, three-phase (16 ampere fla.)

The available service for the above building is 240 volt, three-phase four-wire delta. See Table 8-2 for the calculations. The first step is to take all the lighting and receptacle loads, and all other phase-to-neutral loads, to determine the lighter legs and the neutral. Divide the total amperage obtained by 240. This will yield the amperage on the neutral. The neutral

Table 8-2

		Phase A	Phase B	Phase C	Neutral
Lighting, office	$1000 \times 3.5 \times 1.25 =$	18		18	18
Lighting, manufacturing area	$3000 \times 2 \times 1.25 =$	31		31	31
Receptacles	25×180	19		19	19
1/2 hp blower, 120 volt	4.4	4.4			4.4
3/4 hp blower, 120 volt	6.4			6.4	6.4
1 hp drill press, 120 volt	7.2	7.2			7.2
3 hp, three-phase conveyor	9.6	9.6	9.6	9.6	
5 hp, three-phase air compressor	15.2	15.2	15.2	15.2	
10 hp, three-phase lathe	28	28	28	28	
3 hp, three-phase grinder	9.6	9.6	9.6	9.6	
Heat, three-phase, 10 kW	24	24	24	24	
AC, three-phase, 16 ampere	16 (omit)				
Largest motor	7	7	7	7	
Totals		173	93	168	86

will see only the motor loads of the 110 volt motors; the largest of these motors is 1 horsepower or 864 watts. This will be multiplied by 125%, which increases the neutral size by 216 watts.

Horsepower	Amperage (Table 430.250)
1 5 hp	15.2
1 3 hp	9.6
1 10 hp	28
1 3 hp	9.6
	62.4 amperage

25% of the largest motor is 28 × 0.25 = 7 amperes. This will bring the motor load up to 69.4 amperes.

Now combine the single-phase and three-phase loads as shown in Table 8-2. The lighting loads or 120 volt loads were put on the A phase and C phase because the National Electric Code requires that the high leg be on the B phase (Article 408.3E).

For larger three-phase loads, it may be more desirable to use 480/277 volt systems. The motor and lighting loads would be taken off the 480/277 volt side. Small loads such as receptacles and incandescent lighting would be taken off the low side of a transformer installation.

8-4 THREE-PHASE COMMERCIAL BUILDING, MEDIUM SIZE

A three-story office building contains 21,500 square feet—16,500 square feet of office space and the remaining 5000 square feet common area, such as corridors, restrooms, mechanical rooms, and storerooms. There are

6 lighting standards, each with two 500 watt mercury vapor lamps, 480 volts

1 elevator, 480 volt, three-phase, 15 horsepower, 15 minute rated

125 kW of heating, 480 volt three-phase

100 tons of air conditioning

Determine the size of the service to feed this building. This is a shell calculation and not all the information needed is given. Some estimated values will be needed to finish the calculation since the exact number of re-

Table 8-3

		Phase A	Phase B	Phase C	Neutral
Lighting, office	21500 × 3.5 × 1.25 =	118	118	118	118
Lighting, storage	5000 × 0.25 × 1.25 =	7	7	7	7
Receptacles	21,500 × 1				
First 10 kW at 100%	10,000	13	13	13	13
Remainder at 50%	11500 × 0.5	7	7	7	
125,000 Heat, 3 phase, 460 volts	125,000	157	157	157	
100 tons AC, 124 amperes 3 phase (omit)					
Elevator, 15 hp, 3 phase, 15 minute rated		18	18	18	
Largest motor	5.25	5.25	5.25	5.25	
Totals		325	325	325	125

ceptacles, special lighting, copy machines, and other special circuits can't be determined until the individual tenants start occupying the building, long after the main service has been hooked up.

The receptacles were calculated at $460 \times \sqrt{3} = 796$ volts because that is what the main service will see, even though the receptacles are 120 volts.

The loads in Table 8-3 were split between three phases. Since three phases are available for line-to-neutral loads, the line-to-neutral loads are divided by the phase voltage times the number of phases utilized: $265 \times 3 = 796$. The three-phase portion of the load is the line voltage times $\sqrt{3}$: 460×1.732 ($\sqrt{3}$) = 796.

In a three-phase, four-wire wye system, the line-to-neutral load can be split between three phases instead of two. The line-to-neutral loads are divided by the phase voltage times the number of phases: $120 \times 3 = 360$. The three-phase portion of the load is the line voltage times $\sqrt{3}$: 208×1.732 ($\sqrt{3}$) = 360.

8-5   CONVENIENCE STORE

The next example will be a convenience store. The first calculation (Table 8-4) utilizes a three-phase, four-wire high-leg delta. The second calculation (Table 8-5) utilizes a three-phase, four-wire wye. The convenience store contains the following:

Table 8-4 Convenience store, three-phase, four-wire delta

		Phase A	Phase B	Phase C	Neutral
Lighting	4500 × 3 × 1.25 =	70		70	70
45 Feet show window	45 × 200 × 1.25 =	47		47	47
6 Coolers, 840 watts, 120 volts	840 × 6	21		21	21
2 Coffee makers, 2500 watts, 120 volts	2500 × 2	21		21	21
32 Convience outlets	180 × 32	24		24	24
6 Gas pumps, 120 volt, 4.4 amperes		13		13	13
1 Sign circuit	1200/120 × 1.25 =			13	13
2 Microwaves, 120 volt, 750 watts	750 × 2	6		6	6
6 Display cases, 600 watts, 120 volts	600 × 6	15		15	15
1 Hot-food counter, 240 volt, single phase	12,000	50	50		
1 Gas dispenser, 10 ampere, 120 volts				10	10
				8	8
1 Cash register, 960 watts, 120 volts	10 amps	10			10
2 Compressors, single phase, 240 volts			10	10	
		56	56	56	
2 AC, 28 amperes, 3 phase		7	7	7	
Largest motor 25%			21	21	
1 Water heater, 240 volts, 5000 watts, single phase		340	144	342	258
Totals					

4500 square foot

45 feet of show window

6 coolers, 840 watts, 120 volts each

2 coffee makers, 2500 watts, 120 volts each

32 general-purpose convenience outlets

6 gas pumps, ½ horsepower, 120 volts

1 sign circuit, 1200 watts

6 display cases, 120 volts 600 watts

2 microwaves, 750 watts each, 120 volts

1 hot-food counter, 12 kW, 240 volts, single phase

Table 8-5 Convenience store, three-phase, four-wire wye

			Phase A	Phase B	Phase C	Neutral
Lighting	$4500 \times 3 \times 1.25$	= 16,875	47	47	47	47
Show window	$45 \times 200 \times 1.25$	= 11,250	31	31	31	31
6 Coolers, 840 watts, 120 volts	840×6	= 5040	14	14	14	14
2 Coffee makers, 2500 watts each, 120 volts	2500×2	= 5000		21	21	21
32 convenience outlets	180×32	= 5760	16	16	16	16
6 Gas pumps, 120 volt, 4.4 amperes each	$4.4 \times 6 \times 120$	= 3160	9	9	9	9
1 Sign circuit, 1200 watts, 120 volts	1200×1.25	= 1500	13			13
2 Microwaves, 750 watts each, 120 volts	750×2	= 1500	6	6	6	6
6 Display cases, 600 watts, each 120 volts	600×6	= 3600	15	10	10	10
1 Hot-food counter, 208 volts, single phase, 12 kW	12,000	= 12,000	50		50	
1 Gas dispenser, 10 amps, 120 volts	1200	= 1200		10		10
1 Cash register, 960 watts, 120 volts	960	= 960		8		8
2 Compressors, 1 phase, 208 volts, 10 amperes each	2400×2	= 4800	10	10 10	10	
2 Air conditioners, 32 amperes each, three phase	28×2	= 23,240	64	64	64	
Largest motor 25%		= 2880	8	8	8	
1 Water heater 208 volts, 5000 watts, single phase		= 5000	24	24		
Totals			302	288	286	185

1 gas dispenser, 10 amperes, 120 volts

1 cash register, 960 watts, 120 volts

2 air conditioners, 240 volts, three-phase, 28 amperes each

3 compressors, 10 amperes, 240 volts, three-phase

In this example, a required sign circuit was added (Article 220.14F), which also refers to Article 600.5. Basically, Article 600.5 spells out the requirements for the sign circuit. The Article states that each commercial occupancy accessible to pedestrians shall be provided with at least one outlet in an accessible place at each entrance to each tenant space for a sign or outline lighting by a branch circuit rated at least 20 amperes that supplies no other load. Service hallways or corridors are not be considered accessible to pedestrians.

Also in the example, a show window was calculated at 200 watts per linear foot, according to Article 220.43A. This allows receptacles and power for electrical displays at the show window, which help attract customers.

Each phase has a different amperage on it, including the neutral. Phases A and C are nearly the same. All of the 120 volt and 240 volt three-phase loads are on these phases. Phase B, the high leg, is considerably smaller since only single-phase, 240 volts and three-phase loads can be used on this phase. The neutral shows the imbalance between the two phases (A and C) line-to-neutral loads. When combining loads of different characteristics and different divisors, they must be calculated separately:

Single-phase, 240 volt loads: watts ÷ 240 = amperage

Single-phase, line-to-neutral loads: watts ÷ 240 = amperage

Three-phase loads: watts ÷ 415 (240 × $\sqrt{3}$) = amperage

If the power company allows a three-phase, four wire wye system, the results are slightly different, as shown in Table 8-5.

The three phases are nearly the same amperage. Only the neutral is smaller:

Single phase, 208 volts: wattage ÷ 208 = amperage

120 volt load divided between three phases: 120 × 3 = 360 wattage ÷ 360 = amperage

Three-phase load: wattage ÷ (208 × $\sqrt{3}$) 360 = amperage

The calculations are shown in Table 8-6. The results are:

103,265 ÷ 360 = 287 amperes

55,545 ÷ 360 = 154 amperes

If all of the wattage of this convenience store were added up, the results would be close. The difference between Tables 8-5 and 8-6 is that Table 8-5 is balanced more accurately.

Table 8-6 Convenience store, three-phase, four-wire wye

				Phase	Neutral
Lighting	$4500 \times 3 \times 1.25$	=	16,875	16,875	16,875
Show window	$45 \times 200 \times 1.25$	=	11,250	11,250	11,250
6 Coolers, 840 watts, 120 volts	840×6	=	5040	5040	5040
32 Convenience outlets	180×32	=	5760	5760	5760
2 Coffee makers, 2500 watts each, 120 volts	2500×2	=	5000	5000	5000
6 Gas pumps, 120 volt, 4.4 amperes each	$4.4 \times 6 \times 120$	=	3160	3160	3160
1 Sign circuit, 1200 watts, 120 volts	1200×1.25	=	1500	1500	1500
2 Microwaves, 750 watts each, 120 volts	750×2	=	1500	1500	1500
6 Display cases, 600 watts each, 120 volts	600×6	=	3600	3600	3600
1 Hot-food counter, 208 volts, single phase, 12 kW	12,000	=	12,000	12,000	
1 Gas dispenser, 10 amps, 120 volts	1200	=	1200	1200	1200
1 Cash register, 960 watts, 120 volts	960	=	960	960	960
2 Compressors, single phase, 208 volts, 10 amperes each	2400×2	=	4800	4800	
2 Air conditioners, 32 amperes each, three phase		=	23,240	23,240	
Largest motor 25%		=	2880	2880	
1 Water heater 208 volts, 5000 watts, single phase		=	5000	5000	
Totals		=	103,565	103,565	55,845

8-6 ADDING TO EXISTING SERVICE

If the convenience store exists and a new load is to be added, Article 220.87 can be used. The maximum demand data available for 1 year can be used. This data can be obtained from the power company. They will furnish the highest demand in the previous 12 month period. In the example of the convenience store that we have examined, the maximum demand for the previous 12 month period was found to be 94.3 kW. If a three-phase, four-wire wye system is used, the kilowatts obtained from the power company (94.3 kW) divided by 360 times 125% equals the available service: $94.3 \times 1000 = 94,300$ watts $\times 125\% \div 360 = 327$ amperes. If the size of the service is greater than 327 amperes, the new load could be added to the service. On the other hand, if the service is sized by the load calculation in Section 8-5, which was 312 amperes, the closest wire size in THW copper conductors is 400 Mcm, which is rated for 335 amperes. This would mean that there would only be 8 amperes available before deciding to change the service to accommodate the additional load.

If the service was three-phase, four-wire delta, the maximum demand would be divided by 415 (240 × $\sqrt{3}$): 94,300 ÷ 415 × 125% = 284 amperes. This does not accurately take into account the single-phase 240 volt and 120 volt loads. This means that the A and C phases would show the maximum demand, but the neutral and high leg would not be reflected accurately. Even phases A and C would not be accurate because the single-phase load and line-to-neutral load were lumped together with the three-phase load and divided by 415 instead of 240. The existing neutral in Section 8-4 was sized for 245 amps. This would be a good indication that the phase-to-neutral load is equal to 245 amperes. The high leg in Section 8-4 was sized for 154 amperes. This is also a good indication that the three-phase load would be close to 154 amperes. However, there is some 240 volt single phase load that could be used on the high leg and not on the neutral, which would account for the difference between the high leg and the neutral. In order to size the neutral and high leg it would be possible to use a percentage of the original neutral and high leg to arrive at a consensus. The high leg in Section 8-4 was approximately half of the A and B phases. The neutral is approximately two-thirds the size. If the authority having jurisdiction approves this method, then the calculation could be worked out proportionately. If the service needed to be increased by 50 amperes, the high leg would need to be increased by 25 amperes and the neutral by 36 amperes. However, the load under consideration would have to be evaluated. It may contain only a 120 volt load, or it may only contain a three-phase load. The authority having jurisdiction will provide guidance in this area.

8-7   RESTAURANT CALCULATIONS

A service for a restaurant needs to be calculated. The restaurant has the following equipment (the restaurant is to totally electric):

5000 square feet

2 30 kW ovens, three phase

3 60 ampere fryers, three phase

Electric grill, 50 ampere, three phase

1 Microwave, 1500 watts, 120 volts

1 Dishwasher with booster, 3000 watts, 240 volts

1 Drink dispenser, 1000 watts, 120 volts

1 Coffee maker, 3000 watts, 120 volts

1 Espresso machine, 8000 watts, 240 volts

3 Refrigerators, 8 amperes each, 120 volts

1 Walk-in freezer, three phase, 5000 watts

1 Ice maker, 24 amperes, 240 volts

1 Salad bar, chilled, 7200 watts, single phase, 240 volts

1 Steam table, 50 amperes, three phase

2 Cash registers, 5 amperes, 120 volts

1 Sign circuit, 1200 watts, 120 volts

1 Air conditioning, 54 amperes, three phase

1 Heat, 20,000 watts, three phase

1 Water heater, 40 amperes, three phase

The standard calculation for a restaurant allows one to derate kitchen equipment (Table 220.56). These demand factors can be applied to commercial electric cooking equipment, dishwasher booster heaters, water heaters, and other kitchen equipment that has either thermostatic control or is used intermittently. This demand factor does not apply to electric space heating, ventilating, or air conditioning equipment.

This example (Table 8-7) was set up differently. The total connected load before derating is shown to be 265,055 watts. This is multiplied by the demand factor. The next column is the single-phase and line-to-neutral loads. The next column is the three-phase load. The last column is the neutral load.

If a three-phase, four-wire wye service were available, the load calculation would be balanced on all three phases instead of two (see Table 8-8).

Another way to calculate the restaurant in Section 8-7 is the optional calculation for new restaurants given in Table 220.88. This method uses the total connected load, then, derates it according to the size of the installation. In Table 8-7, the total connected load was 273,166 watts. Table 220.88 is divided into two sections—totally electric restaurants and not-all-electric restaurants. The example in Section 8-6 is totally electric and will be used in this example as well. The first 200 kW is derated to 160 kW and the remainder up to 325 kW is multiplied by 10%. The total connected load is

Table 8-7   Restaurant calculations

				1 Phase	3 Phase	Neutral
Lighting, 5000 square feet	× 2	2 =	10000 × 1.25	12,500		12,500
2 30 kW ovens, three phase	× 2	=	60,000 × 0.65		39,000	
3 three-phase fryers, 60 amperes	× 3	=	74,700 × 0.65		48,555	
Electric grill		=	20,784 × 0.65		13,510	
Microwave		=	1500 × 0.65	975		975
Dishwasher with booster,		×				
single phase		=	3000 × 0.65	1950		
Drink dispenser		=	1000 × 0.65	650		650
Coffee maker		=	3000 × 0.65	1950		1950
Espresso machine,		=	8000 × 0.65	5200		5200
240/120			×			
3 Refrigerators, 120 volts		=	2880 × 0.65	1872		1872
Walk-in freezer three phase		=	5000 × 0.65		3250	
Ice maker, 120 volt		=	5760 × 0.65	3744		3744
Salad bar, chilled,		=	7200 × 0.65		4680	
three phase						
Steam table, three phase		=	20,784 × 0.65		13,510	
2 Cash registers, 120 volts		=	1200 × 0.65	780		780
Sign circuit		=	1200 × 1.25	1500		1500
Air conditioning, three phase,		=	22,447 × 1.25		28,058	
54 ampere						
Heat (omit)						
Water heater, three		=	16,600 × 0.65		10,790	
phase, 40 ampere			265,055			
Totals				31,121	161,353	29,171
				÷ 240	÷ 415	÷ 240
				= 130	= 389	= 122

not adjusted for more than six pieces of equipment, or 125% of the lighting or the largest motor. However the note at the bottom of Table 220.88 instructs us to add both the heat and air conditioning to the calculation. This would increase the connected load by 20,000 watts (for heating):

$$
\begin{array}{r}
285,055 \text{ kW} \\
- 200,000 \qquad = 160,000 \text{ kW} \\
85,055 \times 10\% = \underline{\quad 8506 \quad} \\
168,506 \text{ kW}
\end{array}
$$

If the service were three-phase, four-wire wye, the amperage would be

$$168,506 \text{ kW} \div 360 = 468 \text{ amperes}$$

Table 8-8

		Phase A	Phase B	Phase C	Neutral
Single phase	31,121 ÷ 240 =	130		130	
Three phase	161,353 ÷ 415 =	389	389	389	
Neutral	29,171 ÷ 240 =				122
Total		519	389	519	122

Table 8-9   Restaurant calculations

				Phase A	Phase B	Phase C	Neutral
Lighting	5000 × 2	=	10,000 × 1.25 ÷ 360	35	35	35	35
2 30 kW ovens, three phase	30,000 × 2	=	60,000 × 0.65 ÷ 360	108	108	108	
3 three-phase fryers, 60 amperes	24,900 × 3	=	74,7000 × 0.65 ÷ 360	135	135	135	
Electric grill	20,784	=	20,784 × 0.65 ÷ 360	38	38	38	
Microwave	1500	=	1500 × 0.65 ÷ 120	0	0	8	8
Dishwasher with booster, single phase	3000	=	3000 × 0.65 ÷ 208	9	9	0	0
Drink dispenser	1000	=	1000 × 0.65 ÷ 120	5	0	0	5
Coffee maker	3000	=	3000 × 0.65 ÷ 208	0	9	9	9
Espresso machine, 208/120	8000	=	8000 × 0.65 ÷ 208	25	0	25	25
			× 0.65 ÷ 360	0	0	0	0
3 Refrigerators, 120 volts	2880	=	2880 × 0.65 ÷ 360	5	5	5	5
Walk-in freezer three phase	5000	=	5000 × 0.65 ÷ 360	9	9	9	
Ice maker, 120 volt	5760	=	5760 × 0.65 ÷ 208	0	27	27	0
Salad bar, chilled, three phase	7200	=	7200 × 0.65 ÷ 208	23	23	0	
Steam table, three phase	20,784	=	20,784 × 0.65 ÷ 360	3.5	3.5	0	3.5
2 Cash registers, 120 volts	1200	=	1200 × 0.65 ÷ 240	3	3		3
Sign circuit	1200	=	1200 × 1.25 ÷ 120	0	0	12.5	12.5
Air conditioning, three phase, 54 ampere	22,447	=	22,447 × 1.25 ÷ 360	41	41	41	0
Heat	20,000 (omit)		÷ 360		0	0	0
Water heater, three phase, 40 ampere	16,660	=	16,600 × 0.65 ÷ 360	30	30	30	0
			265,055				
Totals				512	505	512	94

The neutral is not calculated in Table 220.86, but all other optional calculations require that the neutral be calculated by Table 220.61 or by the standard method. The neutral was calculated at 122 amperes. If the service were three-phase, four-wire delta, the amperage would be

$$168,507 \text{ kW} \div 415 = 406 \text{ amperes}$$

This calculation does not give an accurate calculation of the high leg. Since only the three-phase load and single-phase load without a neutral can be used on the high leg, the amperage could be determined proportionately. Since the three-phase load was 86% of the total load, the load could be broken down as follows:

$$168,507 \times 86\% = 144,916 \text{ watts or } 144,916 \div 415 = 349 \text{ amperes}$$

$$168,507 - 144,916 = 23,591 \text{ watts or } 98 \text{ amperes } (23,591 \div 240)$$

Phase A	Phase B	Phase C
349	349	349
98		98
447	349	447

The authority having jurisdiction can give guidance on this calculation. The neutral should be calculated by the standard method, which was 122 amperes.

CHAPTER 8 TEST

1. An office building is 24,000 square feet. If the receptacle load is not known, what is the minimum value that can be used?
2. A small restaurant has four pieces of equipment. What is the demand factor that can be used?
3. What is the minimum lighting load for Problem 1?
4. The portion of the lighting in a hospital calculates to 125,000 watts, excluding operating rooms. What is the minimum watts allowed after the demand factor?
5. What is continuous load?

6. What are the minimum requirements for a sign circuit for a commercial building?

7. A store has 60 feet of store-front window. How much allowance, if any, is needed?

8. What is the minimum lighting allowed for a building that has 1500 square feet of office and 3000 square foot of warehouse?

9. A warehouse is 25,000 square feet. What is the minimum watts allowed for lighting?

10. A 480/277 volt, three-phase, four-wire service is used for a building. The neutral is used for fluorescent fixtures. The lighting calculates out to 280 amperes per phase. What is the minimum load of the neutral?

11. An existing commercial building needs some alterations. The power company has provided the highest reading of the past 12 months. What percentage must be added to that to determine if the existing service is large enough?

12. In a commercial building, what is the watt rating of a 20 ampere general-use receptacle?

13. The optional calculation method for a new restaurant initially yielded 325 kW. The restaurant is totally electric. What would be the derated load in watts?

14. In Problem 13, if the restaurant were not totally electric, what would be the derated load in watts using the optional method?

15. A service requires a 460/277 volt, three-phase, four-wire, 600 ampere service. The neutral calculates to 300 amperes. What size parallel-conductor THW copper would be needed?

16. What size rigid conduit will be needed in Problem 15?

17. What size copper grounding electrode conductor would be needed in Problem 15 to ground the building steel?

18. What size feeder is needed for six elevators rated for continuous duty if each elevator draws 22 amperes at 460 volts?

19. An office building is serviced with a 460/265 volt, three-phase, four-wire service. It is determined that 150 amperes of three-phase, 120 volt are needed for receptacles and other 120 volt loads. What is the amperage of this portion of the load at 460 volts?

20. What would be the overcurrent short circuit and ground-fault protection device for the feeder in Problem 18 using an inverse-time breaker?

21. An office building has 15,000 square feet. There are 125 200 watt fixtures in the lighting design. How many circuits are needed for this project? The lighting is 277 volts.

22. Is the lighting allowance in Problem 21 large enough for the building?

23. How many 200 watt fluorescent lights can be put on a 20 ampere circuit at 277 volts?

24. 15,000 watts are needed for a 120 volt load. If a 240 volt single-phase transformer were used, what would be the amperage of the secondary side of the transformer?

25. If a three-phase transformer were used in Problem 24, what would be the amperage on the secondary side of the transformer?

chapter 9

MISCELLANEOUS SERVICE ENTRANCE CALCULATIONS

9-1 SCHOOL CALCULATION

A school has a 480/277 volt service consisting of the following:

60 classrooms, 36,000 square feet

Hallways and corridors, 6000 square feet

Storage spaces and closets, 3000 square feet

Gym, 12,000 square feet

Office and administration, 5000 square feet

Outside lighting, 480 volt, three-phase, 60,000 watts

Cafeteria, 5400 square feet, with

    2 ovens, 30 kW, three-phase, 208 volts

    1 grill, 45 kW, three-phase, 208 volts

    3 deep fryers, 60 amperes, three-phase, 208 volts

    1 freezer, 30 amperes, three-phase, 208 volts

    3 refrigerators, 1200 watts each, 120 volts

This chapter contains material reprinted with permission from NFPA 70-2005, *National Electrical Code*®, Copyright © 2004, National Fire Protection Association, Quincy, MA 02269. This reprinted material is not the complete and official position of the NFPA on the referenced subject, which is represented only by the standard in its entirety. National Electrical Code and NEC are registered trademarks of the National Fire Protection Association, Quincy, MA.

2 microwaves, 1000 watts each, 120 volts

1 steam table, 30 amperes, 208 volts, three-phase

1 10,000 watt water heater, 208 volts, three-phase

1 dishwasher with booster, 3000 watts, 120 volts

400 general-purpose receptacles throughout the school

Heating consists of 250,000 watts, three-phase, 480 volts

Air conditioning, 361 amperes, 480 volts, three-phase motor

Lighting will be 277 volts, except for the outdoor lighting, which is 480 volts

What size 460/265 volt service is needed for the school? (See Table 9-1.) All of the amperage was converted to watts to make the calculations easier.

The following equipment was converted:

3 deep fryers, 60 amperes, three-phase 208 volts: $(3 \times 60 \times 208 \times 1.732)$ = 64,800 watts

1 freezer, 30 amperes, three-phase, 208 volts: $(30 \times 208 \times 1.732)$ = 10,800 watts

1 steam table, 30 amperes three-phase, 208 volts: $(30 \times 208 \times 1.732)$ = 10,800 watts

1 air conditioner motor, 361 amperes, 460 volts: $(361 \times 460 \times 1.732)$ = 286,995 watts

In column 5 of Table 9-1, the total connected load is calculated before demand factors are applied. Column 6 shows the demand factors that were applied. Lighting was calculated at 125%. The demand factor of 65% was used for more than six pieces of kitchen equipment (see *National Electrical Code®*, Table 220.56). The receptacles were calculated by Table 220.44. The air conditioner was the largest motor and was calculated at 125%.

If this service were calculated by the standard method, the total in column 8 would be 801,306 watts divided by $460 \times 1.732 = 1008$ amperes. The neutral only consists of the 277 volt lighting, which totals 190,063 watts $\div 795$ $(277 \times 3) = 239$ amperes.

Article 220.86 of the *National Electrical Code®* allows a school to be calculated by an optional method. The total connected load before the demand factor (column 5) is divided by the square footage of the school building. The total square footage of the school building is 67,400 square feet:

Table 9-1   School calculation

	1	2	3	4	5
Lighting:					
Classroom	36,000 ×	3 =	108,000 ×	1.25 =	135,000
Hallways & corridors	6000 ×	0.5 =	3000 ×	1.25 =	3750
Storage spaces	3000 ×	0.25 =	750	1.25 =	937.5
Gym	12,000 ×	1 =	1200	1.25 =	15,000
Office	5000 ×	3.5 =	17,500	1.25 =	21,875
Cafeteria	5400 ×	2 =	10,800	1.25 =	13,500
Total square footage of school	67,400			Total lighting	190,062.5
Equipment:					
Grill, 45 kW, 3 phase, 208 volts	45,000 ×		45,000 ×	0.65 =	29,250
2 ovens, 30 kW each, 3 phase, 208 volts	30,000 ×	2 =	60,000	0.65 =	39,000
3 deep fryers, 60 amp each, 3 phase 208 volts	21,600 ×	3 =	64,800	0.65 =	42,120
1 freezer, 30 amp, 3 phase, 208 volts	10,800 ×	1 =	10,800	0.65 =	7020
3 refrigerators, 1200 watts each, 120 volts	1200 ×	3 =	3600	0.65 =	2340
2 microwaves, 1000 watts each, 120 volts	1000 ×	2 =	2000	0.65 =	1300
1 steam table, 30 amp, 208 volt 3 phase	10,800 ×	1 =	10,800	0.65 =	7020
1 10 kW water heater ,3 phase, 208 volts	10,000 ×	1 =	10,000	0.65 =	6500
1 dishwasher with booster, 3000 watts, 120 volts	3000 ×	1 =	3000	0.65 =	1950
400 general purpose outlets	400 ×	180 =	72,000		
1st 10 kW	10,000 ×	100% =		=	10,000
Remainder at 50%	62,000 ×	50%		=	31,000
Heating, 200 kW, 480 vol,t 3 phase	200,000	omit			
Air conditioning ,361 amperes, 460 volt, 3 phase	286,995 ×	1 =	286,995	1.25 =	358,744
Outside lighting	60,000 ×	1 =	60,000 ×	1.25 =	75,000
					729,558
			819,058		

$$819,058 \div 67,400 = 12.15 \text{ watts per square foot}$$

The first 3 watts per square foot are calculated at 100%. Over 3 watts and up to 20 watts per square foot are derated at 75%. Over 20 watts per square foot are derated at 25%

Since this school calculated out at 12.15 or 12 watts per square foot, the size of the service would be:

$$\frac{-3}{9} = \frac{3 \times 67,400}{9 \times 67,400 \times 0.75} = \frac{= 202,200 \text{ watts}}{404,400}$$

12 watts

606,600 watts ÷ 795 (460 × 1.732) = 763 amperes

The neutral would have to be calculated by the standard method, which yielded 239 amperes (total lighting = 190,062 ÷ 795 = 239 amperes).

9-2 COMPUTING FARM LOADS

If a farm has a single service and has two or more buildings with branch circuits besides the dwelling unit, the dwelling unit can be computed by either the standard method or optional method, unless the dwelling has electric heat and the farm has an electric grain-drying system; then the optional method cannot be used. The nondwelling loads having branch circuits or feeders can be computed with the demand factor of Table 220.102, which states that for loads expected to operate without diversity, but not less than 125% of the largest motor, the first 60 amperes is computed at 100%, the next 60 amperes is computed at 50%, and the remainder of the other load is computed at 25%.

Individual loads computed in accordance with Table 220.103 can have a demand factor applied to them as follows:

Largest load = 100%

Second largest load = 75%

Third largest load = 65%

Remaining loads = 50%

The dwelling unit can be added to the total of the above, computed by either the standard or optional calculation method, unless the dwelling has electric heat and the farm has an electric grain drying system. In that case, the optional calculation cannot be used.

9-3 COMPUTING MOBILE HOMES AND MOBILE HOME PARKS

Calculations for a mobile home are covered in Article 550. The service to a mobile home consists of a 50 ampere power supply cord unless the mobile home is equipped with factory-installed gas or oil-fired central heating equipment and cooking appliances, then a 40 ampere power supply cord is permissible. If the computed load is more than 50 amperes, then a weatherproof mast head with four continuous, insulated, color-coded feeder conductors, one of which is the equipment-grounding conductor, or an approved raceway from the disconnect of the mobile home to the underside of the mobile home may be used. There must be a way to attach a properly sized junction box or fitting to the raceway under the mobile home. The manufacturer must provide the size of the feeder conductors and junction box in writing.

To calculate the number of branch circuits needed for lighting and receptacles in the mobile home, the following calculation is needed:

$$\frac{3 \times \text{length} \times \text{width}}{120 \times 15 \text{ (or 20)}}$$

For a mobile home that is 60 feet by 12 feet, determine the number of 20 ampere branch circuits needed:

$$\frac{3 \times 60 \times 12}{120 \times 20} = 0.9 \text{ or 1 circuit}$$

In addition to the one lighting and receptacle circuit, two small appliance circuits are required at 1500 watts each. If a laundry area is provided, a 1500 watt circuit would be required. General appliances include the furnace, water heater, range, and central or room air conditioners.

A mobile home contains the following items. Calculate the service to this unit.

60 by 12 = 720 square feet × 3 watts per square foot = 2160 watts

Heater, 750 watts, 240 volts

Fan, 250 watts, 120 volts

Dishwasher, 600 watts, 120 volts

Range, 7500 watts, 240 volts

60 by 12 = 720 square feet × 3 watts per square foot	= 2160 watts
2 Small appliance circuits, 1500 watts each	= 3000
1 Laundry circuit, 1500 watts	= 1500
	6660 watts

The first 3000 watts are at 100% and the remainder is at 35%:

$$\begin{array}{rl}
6660 & \\
-3000 \times 100\% = & 3000 \\
3660 \times 35\% = & \underline{1281} \\
& 4281 \text{ watts} \div 240 = 18 \text{ amperes}
\end{array}$$

The heater is 750 watts and is calculated at 100%. If air conditioning is present, the smaller of the two loads may be omitted. A blower fan is 250 watts and is multiplied by 125% because it is the largest motor. A dishwasher is 600 watts 120 volts and is added in. A free-standing range is 6500 watts and is calculated by Article 550.18B5 at 80%. A free-standing range would be calculated by 550.18B5, but separate ovens and cooking units would be calculated by adding them together with other fixed appliances (water heater, dishwasher, clothes dryer, or waste disposal). If there are three or more of these appliances, a demand factor of 75% can be applied. Since only one appliance, the dishwasher, is present no deration was possible.

The largest leg in Table 9-2 is 48 amperes, so a 50 ampere power cord will be allowed. If the service were calculated above 50 amperes, power cords could not be used and the minimum service would be 100 amperes.

Services for mobile home parks are calculated by Table 550.31. The load can be calculated on of 16,000 watts per lot or by the largest typical mobile home that is allowed in the park calculated by Article 550.18, whichever is larger.

Table 9-2 Mobile home calculation

			Phase	Phase
Lighting, 60 × 12 = 720 square feet	720 × 3 = 2660			
2 Small appliance circuits	1500 × 2 = 3000			
1 Laundry circuit	1500 × 1 = 1500			
	7160			
	−3000 × 100% = 3000			
	4160 × 35%	1456		
		4156/240 =	18	18
Heater, 240 volt, 750 watt	750	= 750/240 =	2	2
Fan, 250 watts, 120 volts	250 × 1.25	= 313/120	3	
Dishwasher, 600 watts, 120 volts	600	= 450/120 =		4
Range, 6500 watts, 240 volts	6500	× 80% = 5200/240 =	22	22
Total			45	46

Example. If a mobile home park were set up for 15 mobile homes and each unit was set up for 100 amperes, the feeder for these homes would be $100 \times 15 \times 26\%$ (Table 550.31) = 390 amperes.

9-4   RECREATIONAL VEHICLES AND PARKS

Recreational vehicles such as travel trailers, camping trailers, truck campers, and motor homes are intended to provide temporary living quarters for recreation, camping, or travel. They either have their own power or are mounted on or pulled by another vehicle. The rules for recreational vehicles and parks are found in Article 551.

There are four basic electrical hookups for recreational vehicles:

1. 15 ampere power supply. One 15 ampere circuit to supply lights, receptacle outlets, and fixed appliances, 120 volt maximum.

2. 20 ampere power supply. One 20 ampere circuit to supply lights, receptacle outlets, and fixed appliances, 120 volt maximum.

3. 30 ampere power supply. Two to five 15 to 20 ampere circuits to supply lights, receptacle outlets, and fixed appliances, 120 volt maximum. Not more than two thermostatically controlled appliances (air conditioner and water heater) If there are more than two thermostatically controlled appliances, an isolation switch or energy management system must be installed.

4. 50 ampere power supply. For more than five circuits without a listed energy management system, 120/240 volt 50 ampere maximum. The receptacle and attachment plugs are shown in Table 551.46C

Every manufacturer must attach a permanent label to the exterior skin of the vehicle at or near where the power cord enters the vehicle:

THIS CONNECTION IS FOR 110–125 VOLT AC,
60 HZ, _____ AMPERE SUPPLY

or

THIS CONNECTION IS FOR 120/240 VOLT AC,
3 POLE, 4 WIRE, 60 HZ, _____ AMPERE SUPPLY

In either of the above, the blank space is for the correct ampere rating of the unit.

Recreational vehicle parks must have at least one 20 ampere, 125 volt receptacle at each site. There must be a minimum of 5% of sites with 125/250 volt, 50 ampere receptacles. Seventy percent of the sites must be equipped with 125 volt, 30 ampere receptacles.

The calculated load for these sites are as follows:

9600 volt amperes per site for a 50 ampere, 120/240 volt supply.

3600 volt amperes per site for 20 ampere and 30 ampere supplies.

2400 volt amperes per site for a 20 ampere supply.

600 volt amperes per site for 20 ampere supplies dedicated to tent facilities.

The demand factor of Table 551.73 is the minimum allowable demand factor that is allowed in calculating loads for a service and feeders. Where the electrical supply for a recreational vehicle site has more than one receptacle, the calculated load shall only be computed for the highest-rated receptacle. Note that some parks will have a 50 ampere and a 30 ampere receptacle at the same site so they can offer more versatility. In a situation like this, it is only necessary to calculate the larger outlet since only one outlet will be used at one time.

Example. A recreational vehicle park has space for 80 sites. What is the minimum load required?

80 × 5% = four 50 ampere, 120/240 volt sites

80 × 70% = 56 30 ampere, 120 volt sites

Required sites, four 50 ampere 120/240 volt sites and 56 30 ampere 120 volt sites; the remaining 20 sites will consist of 10 dedicated tent sites and 10 20 ampere, 120 volt sites.

$$
\begin{aligned}
4 \times 9600 \quad &= \quad 38{,}400 \text{ watts} \\
56 \times 3600 &= 201{,}600 \text{ watts} \\
10 \times 2400 &= \quad 24{,}000 \text{ watts} \\
10 \times 600 \quad &= \quad \underline{\quad 6000 \text{ watts}} \\
\text{Total} \quad &= 270{,}000 \text{ watts} \times 41\% \ (\text{Table } 551.73) = 110{,}700 \text{ watts} \\
&\quad \div 240 = 461 \text{ amperes}
\end{aligned}
$$

If there are other loads such as a clubhouse, office, and swimming pool, these loads must be calculated separately and added to the site service.

Note: The length of the site feeders may cause voltage drop problems and would have to be taken into account.

The neutral for the 125 volt loads must be the same size as the ungrounded conductor. Recreational site feeder conductors must be adequate for the load supplied and must not be less than 30 amperes.

9-5 HIGH-RISE HOTEL

Some buildings have two or more different occupancies under the same management; for instance, a hotel may have a restaurant, ballroom, bar, conference rooms, gift shops, and beauty shops as well as guest rooms. Each occupancy within the building must be calculated by the rules that govern that occupancy.

Example. A 16 story hotel has the following:

750 guest rooms, each room 450 square feet = 337,500 square feet

Corridors and hall ways in room areas = 13500 square feet

Lobby = 2000 square feet

    8 convenience outlets

Gift shop = 3000 square feet

    20 feet of show window

    1 sign circuit

    12 outlets

House maintenance with laundry, 2500 square feet

    12 washing machines, 12 amperes, 120 volts each

    6 clothes dryers at 6000 watts each, 208 volts, single phase

Office = 2000 square feet

    16 convenience outlets

Restaurant = 3000 square feet

    1 walk-in freezer, 5000 watts, 208 volt, three phase

    1 ice maker, 208 volts, single phase, 24 amperes

    1 salad bar, chilled, single phase, 40 amperes, 208 volts

    1 steam table, three-phase, 208 volts, 50 amperes

    2 cash registers, 5 amperes each, 120 volts

1 juice dispenser, 6 amperes, 120 volts

3 refrigerators, 6 amperes, 120 volts

1 coffee maker, 3000 watts, 120 volts

Conference rooms = 10,000 square feet

4 elevators, 460 volts, 52 amperes each

The service for this building will be 460/277 volts, three-phase, four-wire wye. In Table 9-3, the different occupancies are grouped with their associated equipment within each group. The exception to this is the convenience outlets, which are shown in the specialized equipment section because there are more than 10 kW in convenience outlets throughout the building. The outlets in the guest rooms are not shown because they are part of the lighting and receptacle load and are not calculated separately (Table 220.14J); however, there are 150 convenience outlets in the corridors and hallways of the guest rooms for housekeeping purposes (10 outlets per floor × 15 floors). The continuous-duty rule that applies to the lighting does not apply to the guest rooms. However, in the other occupancies, where the lighting is likely to be on for 3 hours or more, the 125% rule does apply (Table 220.14J).

The lighting in the guest rooms totaled 675000 watts. Table 220.42 allows guest rooms without electric cooking to be derated or adjusted as follows:

First 20,000 watts or less at 50%

From 20,001 watts to 100,000 watts at 40%

Remainder over 100,000 watts at 30%

In Table 9-3 the total lighting for the guest rooms totaled 675,000 watts before derating. The derating allowed is as follows:

$$
\begin{array}{ll}
675,000 \text{ watts} & \\
-20,000 \times 50\% = & 10000 \text{ watts} \\
\overline{655000} & \\
-80,000 \times 40\% = & 32000 \\
\overline{575,000} \times 30\% = & \underline{172500} \\
& 214,000 \text{ watts}
\end{array}
$$

The receptacles throughout the hotel in the different occupancies were grouped together under "specialized equipment" because since there is

Tablle 9-3   Hotel calculation

Guest rooms	450 × 750	2 =	675,000			
			− 20,000 ×	50% =	10,000	
			655,000			
			− 80,000 ×	40% =	32,000	
			575,000 ×	30% =	172,500	
					214,600	
Corridors and hallways, including stairways	900 ×	15 × 0.25 =		× 125% =	4219	
Total					218,719	218,719
Lobby	2000	×	1	× 125% =	2500	
Total					2500	2500
Gift shop	3000	×	3	× 125% =	11,250	
Show window	20	× 200		× 125% =	5000	
Sign circuit	1200	×		× 125% =	1500	
Total					17,750	17,750
House maintenance	2500 ×	×	2	× 125% =	3125	
12 washing machines	12 × 120 ×		12	=	17,280	
6 clothes dryers	6000	×	6	=	36,000	
Total					56,405	56,405
Office	2000	×	3.5	× 125% =	8750	
Restaurant calculation	3000	×	2	× 125% =	7500	
Walk-in freezer, three phase, 5000 watts	5000			× 65% =	3250	
Ice maker, 24 amperes, 208 volts, single phase	24	× 208		× 65% =	3245	
Salad bar, 40 amperes, 208 volts, single phase	40	× 208		× 65% =	5408	
Steam table, 50 amperes, 208 volts, three phase	50	× 208√3		× 65% =	11,700	
Juice dispenser, 6 amperes, 120 volts	6 ×	3 × 120		× 65% =	1404	
3 refrigerators, 6 amperes, 120 volts	6 ×	3 × 120		× 65% =	1404	
Coffee maker, 3000 watts, 120 volts	3000			× 65% =	1950	
Total					35,861	35,861
Conference rooms	10,000	×	1	× 125% =	12,500	12,500

(continued)

Tablle 9-3 *Continued*

Specialized Equipment							
4 Elevators, 3 phase, 460 volt, 52 amperes each	52 ×	4 ×	$460\sqrt{3}$ × 140% ×	85% =	196,778	196,778
Housecleaning outlets in corridors	150		× 180 =	27,000		
Convenience outlets in lobby	8		× 180 =	1440		
Convenience outlets in gift shop	12		× 180 =	2160		
Convenience outlets in office	16		× 180 =	2880		
Convenience outlets in conference rooms	20		× 180 =	3600		
				37,080		
				− 10,000 × 100% =	10,000	
				27,080 × 50% =	13,540	
					23,450	23,540
						564,053

just one service; the receptacles can be derated if the total exceeds 10 kW. If left separately in each occupancy, the total would not be so readily apparent to derating factors.

The elevators under "special equipment" were shown in a simplified manner. If the elevator motors are rated for continuous duty, Table 430.22E shows that the motors would need to be multiplied by 140% each (continuous duty motor used for intermittent duty). Since there are four elevators, there is 85% derating allowed in Table 620.14. If the motors were rated for 5 minutes, 15 minutes, or 30 and 60 minutes, the derating factor would drop from 140% to 85–90% depending on the motor's rating.

The restaurant equipment was derated to 65% since there were more than six pieces of equipment (Table 220.56).

With the information shown, the service for the hotel would be 564,063 watts ÷ 795 (460 × $\sqrt{3}$). = 710 amperes.

9-6 MACHINE SHOP

A machine shop has a 480 volt, three-phase, delta service with the following:

4000 square feet of machine shop area

30 convenience outlets for the manufacturing area

1000 square feet of office space

10 convenience outlets for the office

1 lathe, three-phase, 480 volts, 40 amperes

3 extruding machines, 50 amperes, 480 volts, three phase

2 milling machines, 60 amperes, 480 volts, three phase

3 arc welders, nonmotor, three phase, 480 volts, 40 amperes, 80% duty cycle

1 air compressor, 480 volts, three phase, 30 horsepower

6 poles for outside lighting with two 1000 watt sodium lamps per pole, 480 volts

1 painting and drying booth, 460 volts, 35 amperes, three phase

Air conditioning, 124 amperes, 460 volts

Heavy-duty saw, 460 volts, three phase 36 amperes

3 shearing machines, 26 amperes, 460 volts, three phase

3 presses, three phase, 460 volts, 30 amperes

2 drill presses, 240 volts, three phase

2 overhead cranes, 25 amperes, 460 volts

3 ventilating fans, 15 amperes, 460 volts

In Table 9-4, the air conditioning is the largest motor and is therefore calculated at 125%.

The three arc welders, nonmotor, have a primary amperage of 40 amperes each. Since the duty cycle is 80%, the multiplying factor is 89% (Table 630.11A). The first two welders are calculated at 100% and the third welder is calculated at 85%:

$$40 \times 89\% \times 100\% \times 2 = 71 \text{ amperes}$$
$$40 \times 89\% \times 85\% \times 1 = \underline{30} \text{ amperes}$$
$$101 \text{ amperes}$$

There are two overhead cranes in the building. Article 610 covers overhead cranes and Table 610.14E shows the demand factor for more than one crane. In this case, the demand factor for two cranes added together is multiplied by 95%.

All of the amperage was converted to wattage to add the total up easier. The total service calculated out to 782,938 watts. Divide this amount by 795 ($460\sqrt{3}$) and the total 460 volt service is calculated. Since the service

Table 9-4   Machine shop, industrial

Manufacturing area	4000 ×	2	× 125%	10,000
outlets for manufacturing	180 ×	30		5400
Office	1000 ×	3.5	× 125%	4375
convenience outlets	180 ×	10		1800
Outside Lighting	6 ×	2000	× 125%	15,000
Air compressor 30 horsepower, 460 volts	40	× 460 × $\sqrt{3}$		31,800
2 milling machines	60 ×	2 × 460 × $\sqrt{3}$		95,400
1 lathe, three phase, 460 volts, 40 amperes	40	× 460 × $\sqrt{3}$		31,800
3 arc welders, 460 volts, three phase, 40 amperes, 80% duty cycle	101	460 × $\sqrt{3}$		80,295
3 extruding machines, 460 volts, 50 amperes each, three phase	50 ×	3 × 460 × $\sqrt{3}$		119,250
Painting and drying booth, 460 volts, three phase, 35 amperes	35	× 460 × $\sqrt{3}$		27,825
Air conditioning, three phase, 460 volts, 124 amperes	124	× 460 × $\sqrt{3}$ × 125%		123,225
Heavy-duty saw, 460 volts, three phase, 36 amperes	36			
3 shearing machines, 460 volts, three phase, 32 amperes	32 ×	3 460 × $\sqrt{3}$		76,320
3 presses, 26 amperes, 460 volts	26 ×	3 460 × $\sqrt{3}$		62,010
2 drill presses, 30 ampere, 240 volt, three phase	30 ×	2 × 240 × $\sqrt{3}$		24,900
2 overhead cranes, 460 volts, three phase, 25 amperes	25 ×	2 × 460 × $\sqrt{3}$ × 95%		37,763
3 ventilating fans, 460 volts, three phase, 15 amperes	15 ×	3 × 460 × $\sqrt{3}$		35,775
Total wattage				782,938
Total amperage				985

is a 460 volt delta, there is no 277 volt neutral. The office lighting would be best served by 120 volt fixtures, but the manufacturing area can be serviced by 460 volt high pressure sodium fixtures.

CHAPTER 9 TEST

1. If a school has an initial load of 2,500,000 watts and is 50,000 square feet, what is the derated load by the optional method in watts?

2. If the service were 460/277, volt, three-phase, four-wire, what would the amperage be in Problem 1?

3. A farm has six buildings, including a dwelling. The dwelling is calculated at 26,000 watts. Buildings 1 and 2 have a 150 ampere combined load but perform the same function. Building 3 has an 80 ampere load, Building 4 has a 50 ampere load, and Building 5 has a 30 ampere load. What is the total load for the service?

4. A 50 by 8 foot mobile home has a 6000 watt free standing stove, a 1500 watt water heater 120 volt, and a 250 watt 120 volt blower fan. What size service is needed for this unit?

5. A nameplate is required to be on the outside, adjacent to the feeder assembly. What is the purpose of this nameplate?

6. What is the number of appliances required before a 75% demand factor can be used?

7. What are the minimum and maximum lengths of a power supply cord?

8. What is the rating used in the service of a 12 kW stove in a mobile home?

9. A mobile home calculates at 63 amperes on the highest phase. What is the minimum size of the service for this unit?

10. Six mobile homes are in a park. The typical home in this park is 16,000 watts. What is the size of the feeder in amperes servicing these units?

11. What is the voltage to a recreational vehicle?

12. What are the sizes of power outlets in a recreational vehicle park?

13. A recreational vehicle park has 100 spaces. How many spaces are required to be 50 ampere? 30 ampere?

14. In Problem 13, how many sites are required to be dedicated tent sites with 15 and 20 ampere electrical supply?

15. What is the calculated value of a 50 ampere site?

16. A recreational park has 50 spaces; 35 are 30 ampere, five are 50 ampere, five are 20 ampere, and five are dedicated tent sites. What is the amperage of the feeder of this park?

17. A hotel has 150,000 square feet of guest rooms without electric cooking. What is the lighting demand?

18. What is the demand factor for a service that has four overhead cranes?

19. In Table 9-4 how many 20 ampere circuits are needed for the machine shop and office lighting at 277 volts?

20. How many circuits are required for the outside lighting in Table 9-4 at 277 volts?

21. How many circuits would be required for the guest rooms of problem 17? Use 120/208 volts.

22. In Table 9-3, what size 480 volt feeder is required for the house maintenance area?

23. In Table 9-3, the restaurant requires what size panel if all of the restaurant were in 120/208 volts?

24. In Table 9-4, what size feeder in THW copper conductors would supply the air conditioner?

25. What size overcurrent device is required in Problem 24?

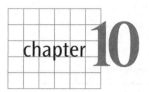

chapter 10

CONDUIT BENDING

10-1   OFFSETS

Conduit bending is an art aided by math. When bending any offset, no matter what the degrees, a right triangle is created. A 30 degree offset with a 6″ rise, for example, has a distance of twice the rise between the marks. Place two marks 12″ apart, and put the hand-bender arrow on the first mark. Pull the bender handle up. Usually, when the bender handle is straight up, the bend is at 30 degrees. Turn the conduit over by 180 degrees and slide the bender down to the second mark. Pull the bender up until the bent leg of the conduit is level. Measure from the floor to the bottom of the conduit, the measurement should be 6″ (see Figure 10-1).

Method 1

Where the two dotted lines join together (Figure 10-1), a right angle is formed. Once again Pythagoras' Theorem comes into play: $A^2 + B^2 = C^2$ (see Figure 10-2).

The height of the offset is side B or 6″. The marks on the conduit are 2 × 6 = 12″ or the hypotenuse, side C. Side A is the third side of the triangle. Normally, we don't pay attention to side A because if we know the height and the angle, we know how far apart our marks need to be, so side A becomes moot. However, occasionally, the length of side A is needed to determine the shrinkage. Shrinkage due to the bending results in a loss of conduit.

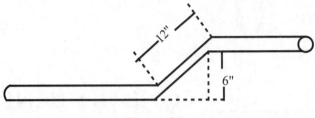

Figure 10-1

In the previous example of a 30 degree offset with a 6″ rise, one side, B, is 6″ and side C, the hypotenuse, is $2 \times 6 = 12″$. Set up the formula of $A^2 + B^2 = C^2$. Then

$$C^2 - B^2 = A^2$$

$$12^2 - 6^2 = A^2 = 144 - 36 = A^2$$

$$A^2 = 108$$

$$\sqrt{108} = 10.39$$

Now subtract 10.39″ from 12: $12 - 10.39 = 1.607″$.

After the bend, the conduit will be 1.607″ shorter than the straight piece you started with before the bend. This can be very important when bending rigid conduit. It is called the developed length. The ability to predetermine the length of a conduit before bending can save time and aggravation. Having to cut and thread a short piece of conduit with an offset can be a hassle. Many times, it needs to be threaded by hand.

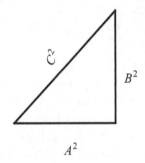

Figure 10-2

Method 2

Method 2 employs a scientific calculator. To find the multiplier for a 30 degree bend, press 30 on the calculator. Next press the sin button. The display will read .5. Next press $\frac{1}{x}$ The display will change to 2. This is the multiplier. Now press the multiplying key, then the height of your offset, in this case 6. The answer will be 12:

$$\boxed{30} \; + \; \boxed{SIN} \; + \; \boxed{\dfrac{1}{X}} \; + \; \boxed{*} \; + \; \boxed{6} \; = \; \boxed{12}$$

This is the distance between the bends. Put 12 in the plus memory.

Now put 30 in the calculator display again. Press tan. The display will change to .5773. Now press $\frac{1}{x}$ and the display will change to 1.732. Press the multiplier key and then enter the height of the offset, 6. The display will change to 10.39.

$$\boxed{TAN} \; + \; \boxed{\dfrac{1}{X}} \; + \; \boxed{*} \; + \; \boxed{6} \; = \; \boxed{10.39}$$

Press the minus memory key and then press memory recall. The display will be 1.607. This will be the shrinkage of a 30 degree offset with a 6″ rise.

Method 3

Method 3 is simply going to the Table 10-1 and choosing the degree of bend that you want to use in column A. For instance, a 15 degree bend with a four inch offset is required. Find 15 degrees in column A. Column B shows the cosecant of the 15 degree angle: 3.864. Multiply 3.864 times the depth of the offset, in this case 4, and the answer is 15.456 inches. This is the distance that is marked for the two bends. Now go to column D, across from 15 degrees. This is the cosecant less the cotangent. The answer is 0.1319 inches

Multiply 0.1319×4 to get the shrinkage of this offset. The answer is 0.5276, or a little more than 0.5 inches.

All three of the methods discussed above will result in the same answer.

Sometimes a mechanic gets caught in the field without a calculator or a chart. Then a fourth method can be utilized to determine shrinkage. Although this method is not as accurate, it will give you satisfactory results.

Table 10-1 Cosecants and Cotangents

A Degrees	B Cosecant of Angle	C Cotangent of Angle	D Shrink per Inch of Rise
1	57.299	57.29	0.009
2	28.6537	28.636	0.0177
3	19.107	19.081	0.026
4	14.336	14.301	0.035
5	11.474	11.43	0.044
6	9.567	9.5144	0.0526
7	8.202	8.1443	0.0577
8	7.185	7.1154	0.0696
9	6.392	6.3138	0.0782
10	5.759	5.6713	0.0877
11	5.241	5.1446	0.0964
12	4.81	4.7046	0.1054
13	4.5	4.3315	0.1685
14	4.134	4.0108	0.1232
15	3.864	3.7321	0.1319
16	3.628	3.4874	0.1406
17	3.42	3.2709	0.1491
18	3.236	3.0777	0.1583
19	3.072	2.9042	0.1678
20	2.924	2.7475	0.1765
21	2.79	2.6051	0.1849
22	2.674	2.4751	0.1989
23	2.554	2.3559	0.1981
24	2.459	2.246	0.213
25	2.366	2.1445	0.2215
26	2.281	2.0503	0.2307
27	2.203	1.9626	0.2404
28	2.13	1.8807	0.2493
29	2.063	1.804	0.259
30	2	1.7321	0.2679
31	1.942	1.6643	0.2777
32	1.887	1.6003	0.2867
33	1.836	1.5399	0.2961
34	1.788	1.4826	0.3054
35	1.743	1.4281	0.3149
36	1.701	1.3764	0.3246
37	1.662	1.327	0.335
38	1.624	1.2799	0.3441
39	1.589	1.2349	0.3541
40	1.556	1.1918	0.3642
41	1.527	1.1504	0.3766
42	1.494	1.1108	0.3832
43	1.466	1.0724	0.3936
44	1.44	1.0355	0.4045
45	1.414	1	0.414

First, lay out on a piece of cardboard or paper the height of the offset. Next place the edge of the ruler on this mark diagonally down to the edge of the paper, the distance of the offset or cosecant. With a second ruler, measure from the end of the cosecant to the opposite end, forming a triangle. Using a 30 degree offset with a 4 inch rise, the cosecant will be 4 × 2 or 8 inches (see Figure 10-3).

At point C, the ruler overhangs the mark by one inch. This is the shrinkage. It should be noted that the cosecant decreases as the angle increases. As the angle increases, the shrinkage per inch increases. Sometimes, the bender will not allow you to bend the second half of the bend because the first bend won't clear the shoe of the bender. This is especially true when using a Chicago Bender or hydraulic bender. If you use fewer degrees, the cosecant per inch is longer and this gives you more room to make the bend. Another point that should be noted is that the higher the rise, the more the distance between bends will be shortened with a larger angle. The cosecant per inch is shorter. For example, if a 15″ rise is needed, a 30 degree bend would have a cosecant of 2 × 15 = 30 inches. A 45 degree bend would have a cosecant of 1.414 × 15 = 21.21 inches.

The larger 45 degree bend is more compact (45 degrees), only taking 21.21 inches instead of 30 inches. Care should be exercised in using larger bends. A 45 degree offset consists of two 45 degree bends, which is 90 degrees total. This is significant since a conduit run is limited to 360 degrees total by the National Electrical Code (NEC) and some job specifications allow only 270 degrees in total bends. This can be solved by putting junction boxes or pull boxes in these runs. Each time a junction box is inserted, the 360 degree or 270 degree rule starts over. Planning the conduit run ahead of time pays big dividends in the end.

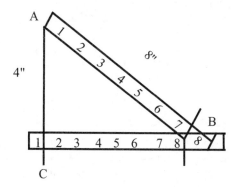

Figure 10-3

Smaller degree offsets use less degrees in the overall count:

10 degree offset	2 × 10 degree = 20 degrees
15 degree offset	2 × 15 degree = 30 degrees
20 degree offset	2 × 20 degree = 40 degrees
25 degree offset	2 × 25 degree = 50 degrees
30 degree offset	2 × 30 degree = 60 degrees
45 degree offset	2 × 45 degree = 90 degrees

Sometimes, but not always, a conduit run will start off with a 90 degree bend out of a panel. The number of offsets and the sizes of the offsets can have a significant impact on the conduit run.

Sometimes a trigonometric table does not contain a column for Cosecant or Cotangent functions. To find a cosecant of a 30 degree angle, find the sin of 30 degrees in the table, which is 0.5. Divide the sin or 0.5 into 1 and the cosecant will be the answer: $1/0.5 = 2$. This works for all angles. The sin of a 15 degree angle is 0.2588: $1/0.2588 = 3.864$. The cosecant of the angle is the multiplier times the depth of the offset.

The same operation can be done to find the cotangent. First find the tangent of the 30 degree angle, which is 0.5774. Divide this into 1 and the answer is the cotangent: $1/0.5774 = 1.732$. Again, multiply the cotangent times the depth of the offset, then subtract the result from the cosecant times the depth of the offset:

$$30 \text{ degree angle cosecant} = 2 \times 6 = 12$$
$$30 \text{ degree angle cotangent} = 1.732 \times 6 = \underline{10.39}$$
$$1.61 \text{ inches}$$

If it is easier, subtract the cotangent from the cosecant, then multiply the depth of the offset: $2 - 1.732 = 0.268 \times 6 = 1.61$ inches.

10-2   MULTIPLE OFFSETS

On occasion, it is necessary to parallel several offsets side by side. If you make four identical offsets and lay them side by side, the spacing won't quite match up from end to end. It is similar to making four identical 90 degree bends and trying to match them up. The spacing just doesn't work out. Multiple 90 degree bends will be covered later in this chapter under segment bending (Section 10-7).

Multiple offsets or parallel offsets should fit together and have the same spacing from top to bottom. If two identical offsets are made and laid down side be side, they do not maintain the same spacing throughout. If one of the offsets is moved until the two offsets do maintain the same spacing throughout, the top and bottom no longer match up. In order to get the offsets to match up in length as well as in spacing, use the following formula: ½ × tangent of angle, times spacing between conduits, plus the outer diameter of the conduit.

Example. Three ½ inch conduits are spaced 2 inches apart and are paralleled with a 30 degree offset of 4 inches. The first bend is 6 inches from the end of the conduit. Where is the starting point of the other two matching conduits?

Conduit B: ½ tangent 30 × 2 + 0.707 = 1.28 inches The starting point of conduit B is the starting point of conduit A plus 1.28 inches, or 7.28 inches. The second bend will be measured from the first bend, as was done on conduit A.

Conduit C: The starting point of conduit C is the starting point of conduit B plus 1.28 inches: 7.28 + 0.72 = 8.56 inches. The second mark of the bend will be made from the first mark of conduit C.

Another way to make multiple offsets that match up is to bend the first offset and lay it down on a piece of cardboard or paper. Make an outline of the offset. Next, lay down the offset that you made next to the outline you just drew. Line up the offset with the drawing, maintaining equal spacing from top to bottom. Notice that there is a gap between the bottom of the offset and the bottom of the drawing. Measure the distance and add that to the next offset.

For example, four ½″ conduits spaced 2″ apart with a 30 degree offset of 3″. Bend the first offset, starting the bend 3″ from the end of the conduit. Make the second bend as you normally would. Lay down the completed offset on a piece of cardboard or paper and make an outline of the offset. Lay the offset next to the outline and get the proper spacing. When the spacing is even from top to bottom, you will notice a gap between the bottom of the outline and the bottom of the conduit.

Figure 10-4A shows the offset with the outline around it. Figure 10-4B shows the offset and the outline even at the bottom. Even though 2″ spacing can be obtained at the bottom and top of the offset, notice how the middle is less than 2″—it is pushed together. Figure 10-4C shows the offset slightly higher than the outline—2″ spacing is maintained from top to

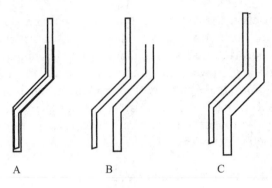

Figure 10-4

bottom. In this example, the gap between the bottom of the offset and the bottom of the outline is 1.28″.

The bend on the second offset will start 1.28″ higher than the first. The first bend of the first offset started at 3″ from the end of the conduit. The first bend of the second offset will start 4.28″ from the end. The second bend will be laid out using the first bend. Continue to add 1.28″ to each offset as you progress. The first bend on the third conduit will start at 5.56″ from the end of the conduit; continue the second bend as a normal offset. The first bend on the fourth conduit will start 6.84″ from the end of the conduit; the second bend will continue as a normal offset. If the conduits started at the same length before bending, the ends will match up just fine.

10-3 KICKS

A kick is half of an offset. It allows you to go left, right, up, or down from the direction in which you are going. Kicks are especially popular just before a 90 degree bend. For instance, a conduit is running along the ceiling and 90s to the left. Immediately after the 90, a 3″ change in elevation downward is needed to continue the conduit run.

Most electrician's will eyeball where the kick should be and make the bend by raising the 90 off the ground by measuring it with a ruler. If you measure back a specific distance, say 12 inches, you now know the length of two of the sides of a right triangle formed by the dotted line in the Figure 10-5. This will allow you to calculate the shrinkage of the kick, which will shorten your conduit a little.

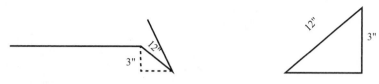

Figure 10-5

The two known of the triangle have are the hypotenuse and opposite sides. The hypotenuse is 12″ and the opposite side is 3″. The unknown side is the adjacent side. In order to find the unknown side, use the formula

$$A^2 + B^2 = C^2 \qquad B = 3; C = 12'' \qquad 12^2 - 3^2 = 135 = \sqrt{135} = 11.62$$

Subtract the adjacent side from the hypotenuse to give the shrinkage: 0.38 inches. By being able to calculate the shrinkage, the almost perfect fit becomes a perfect fit. If it is necessary to find the degrees of an angle, merely divide two sides:

$$\sin = \frac{O}{H} \qquad \cos = \frac{A}{H} \qquad \tan = \frac{O}{A} \qquad \sin = \frac{3}{12} = 0.25$$

$$\cos = \frac{11.62}{12} = 0.9683 \qquad \tan = \frac{3}{11.62} = 0.2582$$

where o = opposite side, H = the hypotenuse, and A = the adjacent side. Look up tables the closest number to 0.25 in the trigonometric under the sin column. It is between 14 and 15 degrees. All three angles are close to 14.5 degrees, which is the angle of the kick. This formula is useful when you need to find the angle of any bend, including those of existing conduit that you are trying to match.

10-4   THREE-POINT SADDLE

An obstruction in the conduit route is not uncommon. An offset will get by the object, but a second offset is needed to get back to the same plane as before. For an obstruction of 4″ in diameter or less, a three-point saddle will work nicely (see Figure 10-6).

To make a three-point saddle, measure to the center of the obstruction. Mark on the conduit where the center of the obstruction would be. If you

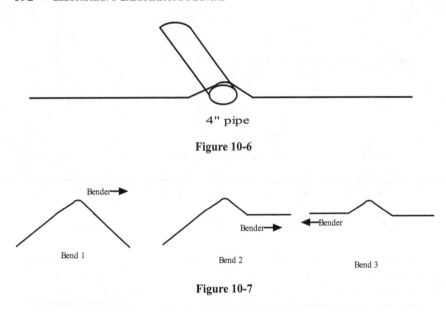

4" pipe

Figure 10-6

Bender➤

Bend 1

Bender➤   ◄─Bender

Bend 2

Bend 3

Figure 10-7

are using a bender that marks the center of the bend, line the mark you made with this mark on the bender. If you do not have this mark on the bender, you will need to find it when you make the bend. It will be necessary to adjust the mark on the conduit to compensate for the difference. Multiply the height of the obstruction times 2.5. In this example the obstruction is a four inch water pipe so $4 \times 2.5 = 10$ inches.

Mark the conduit 10″ above the mark made for the center of the obstruction. Make another mark 10″ on the other side of the center of the obstruction. Now the three bends are marked. Bend a 45 degree bend on the center mark. Turn the conduit over and slide the bender down to the second mark with the bender facing in the same direction. Now bend a 22.5 degree bend. Remove the bender and move to the other side of the two bends where the third mark is. The first two bends are in the same direction; however, the third bend is in the opposite direction. With the conduit facing the same direction as the second bend, make a second 22.5 degree bend. Now place the saddle over the obstruction and it should fit perfectly (see Figure 10.7).

10-5   FOUR-POINT SADDLES

When larger obstructions are encountered, usually larger than 4″ in either depth or width, a four point saddle is more desirable (see Figure 10-8). A

Figure 10-8

four-point saddle consists of two complete equal offsets. The first offset raises the conduit above the obstruction and the second offset brings the conduit back to the original level that it was at. If the bends clear the obstruction evenly, it will look uniform and deliberate. The depth and width of the obstruction will have a significant bearing on the placement of the offsets.

The first thing to do is to measure the center of the conduit to the center of the obstruction and mark it on the conduit (see Figures 10-9 and 10-10). Then mark the beginning of the offset on either side of the center line on the conduit, at points B and C; in this case, 8″. The obstruction is 6″ high and in this case a 30 degree bend will work nicely, so mark out 12″ from mark B and mark C.

Place the bender at point B, pointed toward point D, and bend a 30 degree bend (see Figure 10-11). Slide the bender down to point D, rotate the conduit 180 degrees and bend 30 degrees in the opposite direction. Now

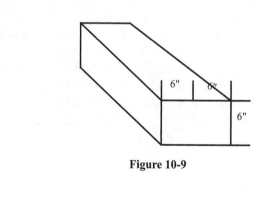

Figure 10-9

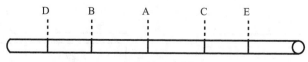

Figure 10-10

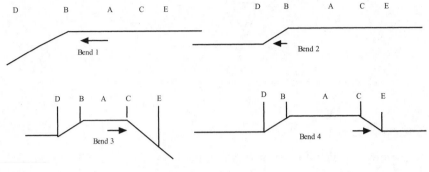

Figure 10-11

place the bender on point C facing the front of the bender towards point E; be sure to rotate the conduit back 180 degrees. Now bend a 30 degree bend. Slide the bender down to point E, rotate the conduit 180 degrees, and bend the fourth 30 degree bend.

10-6 NINETY DEGREE BENDS

Ninety degree bends are an important part of conduit bending. A ninety degree bend is commonly the largest bend you will most likely bend. The characteristics of a 90 are different. With offsets and kicks, the conduit was shorter when the bend was completed. In the case of the 90, a gain is realized. The curve of the of the conduit actually gains in length. The sum of the stub and tail is greater than the length of conduit that you started with. It is always a good idea to bend a 90 with a scrap piece of conduit. Always measure the conduit before bending. Each bender will have its own characteristics. In the case shown in Figure 10-12, a ½" bender has a 5" deduction for the stub. This means that if you want the stub to be 12", you would deduct 5" and mark the conduit 7" from the end. Some mechanics will start

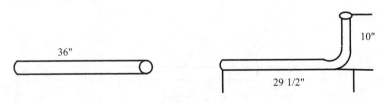

Figure 10-12

the measurement 5″ beyond the end of the conduit and mark the conduit at 12″. This saves the step of subtracting the deduction from the length of the stub. The overall length of the conduit is 36″. Now bend the conduit.

Now add the stub and tail together: 10 + 29½ = 39½″. The sum of the stub plus the tail is 3½″ longer than the straight piece of the conduit that you started with. It should be noted that the size of the conduit as well as the type of bender will affect the amount of gain.

Once the gain and the stub length have been determined, you are now ready to bend 90 degree bends, knowing the length of the stub. A 15″ stub is desired and we determined that a 5″ deduction equals 10″ mark on the conduit. Now, bend the conduit and measure the stub. This will give you the desired 15″ stub.

Back-to-Back Ninety Degree Bends

Often, a back-to-back 90 is needed. There are two ways to bend this: straight push through or reverse bending. In order to use the push through method, there must be enough tail to bend on, or in the case of a Chicago bender, to hit the back roller when the bend is complete. For example, the stub is 15″ with 36″ between the 90's and a 36″ tail (see Figure 10-13). The length of conduit required for this would be 15 + 36 + 36 − 2 × gain (or 2 × 3½) = 80″. Mark off the first bend by subtracting the deduction from the stub (15 − 5 = 10), then make the bend. The deduction is added in front of the bender, making a 15″ stub. The gain goes behind the bend, adding 3½″ to the tail.

Mark the second bend, 36″ − 5″ = 31″, and bend. Since this bend was a push through, the deduction is added in front of the bend, giving 36″ between bends. Before bending and after making a mark at 31″, there is

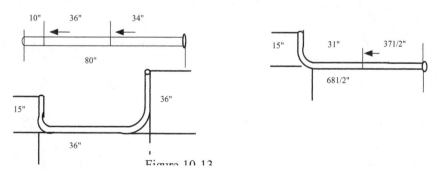

Figure 10-13

37½" of straight conduit after that mark. After making the bend in the same direction as the first bend, the 5" deduction goes in front of the bender to add to the 31". This takes 5" off of the 37½", leaving 32½" plus the gain of the second bend of 3½" or 36".

There are times when the tail is too short to be bent by the push through method and the 90 must be bent in reverse. For example, consider a 15" by 36" by 10" bend (see Figure 10-14). The length of the conduit would be 15 + 36 + 10 − 2 × gain (or 2 × 3½) = 61 − 7 = 54". Mark the first bend as 15 − 5 = 10. Mark the second bend from the other end of the conduit for the stub, less the deduction.

The 5" deduction adds to the front of the bend, giving a 10" stub. The gain always comes behind the bender. After the 5" deduction is marked, there is 38" of straight conduit left. After the bend is made, the 5" deduction adds to the stub (5" + 5" = 10"). The original 38" of straight conduit now becomes 33" plus the gain of 3", or 36".

Sometimes, it is necessary to put three bends in a conduit. The same rules as above apply. Knowing where and how much the deductions and gains will add or subtract will determine where the conduit is marked.

The following series of bends will illustrate this (see Figure 10-15). The first 90 will have a 15" stub down. There will be 48" between the first 90 and second 90, which will turn left for 24", then 90 down for 10". First, determine the length of conduit by adding all of the legs together less the gains from the three 90s: 15 + 48 + 24 + 10 − (3 × 3½ = 10½) = 86½". Now mark the conduit for the first bend: 15 − 5 = 10". From the 10" mark, measure 43" and mark. This will be the mark for the second bend. Now, since the tail will be too short to bend in the same direction, measure the third bend 5" from the opposite end. The conduit will look like the one shown in Figure 10-15.

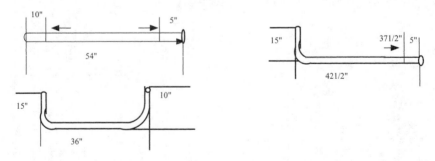

Figure 10-14

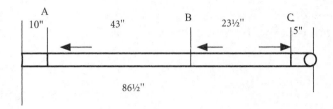

Figure 10-15

Now, bend the conduit at points A and B in the same direction and at point C in the opposite direction. Once the first bend is made, the mark at point B is 43″, which accounts for the 5″ deduction needed for the second bend. Be careful to keep the planes of the bend level and in the right direction (see Figure 10-16).

Different size conduits will have different deductions. This is because different size bender shoes have different radii. A Chicago bender will have a different radius for a half inch conduit than a hand bender. Another point that must be noted is spring back, especially on rigid conduit. This varies between benders, but can be as much as 5 degrees. Spring back is nothing more than the resistance of the conduit to the bend. When you bend a 90 degree bend and either the indicator on the machine or a protractor shows 90 degrees when the bender is released, the conduit will spring back between one and five degrees, leaving an 85 to 89 degree bend. To compensate for spring back, the conduit must be overbent so that when the shoe is released, the conduit springs back to 90 degrees. Often, mechanics will mark the indicator to show spring back on a particular machine. Other bends less than 90 degrees will have spring back as

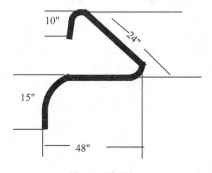

Figure 10-16

well, although not as much. It is a good idea to use a protractor to check the first few bends with an unfamiliar machine.

10-7   SEGMENT BENDING

Although the math involved in segment bending is intensive, the results are spectacular to the trained mechanic. Consider four 1½" conduits with 4" spacing. The assignment is to lay out and bend four 90 degree bends, maintaining 4" spacing throughout and having the couplings match up at both ends. We will need to find the radius and the developed length of each conduit. To start with, find the minimum bending radius—eight times the diameter of the conduit pluse half the outside diameter of the conduit:

$$8 \times 1\frac{1}{2} + \frac{1}{2} \text{ of } 1.90 = 12 + 0.95 = 12.95$$

Once the radius is determined for the inside conduit, add the spacing to each of the successive conduits:

Radius of conduit # 2 = Radius of conduit # 1 + 4" = 16.95"

Radius of conduit # 3 = Radius of conduit # 2 + 4" = 20.95"

Radius of conduit # 4 = Radius of conduit # 3 + 4" = 24.95"

To find the developed length , multiply each radius by ½ π or 1.57:

Developed length for conduit # 1 = 12.95 × 1.57 = 20.33"

Developed length for conduit # 2 = 16.95 × 1.57 = 26.61"

Developed length of conduit # 3 = 20.95 × 1.57 = 32.89"

Developed length of conduit # 4 = 24.95 × 1.57 = 39.17"

In order to join the existing conduit run, we need a 90 with a 70" stub. To determine the length use the following formula: 120 − 70 -39.17 + 49.9 = 60.73". This is a variation of a formula that we will study in greater detail later in this chapter. For now, take a full piece of conduit minus the stub-up minus the developed length plus 2 × the radius. The formula we will be using will be stub-up + tail + developed length − 2 × radius = conduit length. Stub-up = 70, tail = 60.75, developed length = 39.17, 2 × radius = 49.90, so

$$70 + 60.75 + 39.17 - 49.9 = 120''$$

To find out where to mark the conduit to start the bend, subtract the radius from the stub-up:

$$70 - 24.95 = 45.05''$$

From this mark, measure out the developed length of 39.17″ and mark it. From the second mark of the developed length to the end of the conduit will be 35¹³/₁₆″. This can also be determined by subtracting the radius from the tail. 60.75 − 24.95 = 35.8″. The conduit should look like the one shown in Figure 10-17.

The two marks you made will be the beginning mark and end mark of the developed length of 3.17″. Divide 3.17 by a number that will go into 90. The most common are:

1. 90/18 = 5 (18 equal segments of 5 degrees each)
2. 90/30 = 3 (30 equal segments of 3 degrees each)
3. 90/36 = 2.5 (36 equal segments of 2.5 degrees each)

The deciding factor will be how far apart the segments will be. If they are too close together, there will be a problem. If they are too far apart, the conduit will look crimped. The best results occur between ¾″ up to 1¾″ apart. If the developed length is divided by 18, 30, and 36, respectfully, the spacing will be as follows:

39.17/18 = 2.18 inches
39.17/30 = 1.30 inches
39.17/36 = 1.09 inches

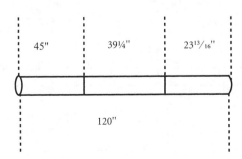

Figure 10-17

1.09" is between $1\frac{1}{16}$ and $1\frac{1}{18}$ inches. There are several ways to mark the conduit. If you have surgical tubing for each of 18", 30", and 36" segments, take the 36" segment tube out and put the first mark of the tubing on the first mark of the developed length. Stretch the tube out until the last mark on the tubing matches up with the end of the developed length on the conduit. The tubing will stretch out evenly, and you can mark the conduit at each mark on the tubing (see Figure 10-18). Now bend each segment 2.5 degrees, the result will give you the segmented 90 degree bend with a 70" stub and a 60.75" tail (see Figure 10-19).

The second way to measure is to use a ruler and measure off $1\frac{1}{16}$" marks between the marks you made for the developed length of your conduit. The next conduit will fit inside the first conduit, so it will be shorter. Conduit #3 will be 4" less on each end, so the stub-up will be 66" and the tail will be 56.75". The developed length will be 4" shorter as well: 20.95" × 1.57 = 32.89". Each successive conduit will be 4" shorter:

Conduit	Stub	Tail	Radius	Developed length, Radius × 1.57	2 × Radius
Conduit #1	58"	48.75"	12.95"	20.33"	25.9"
Conduit #2	62"	52.75"	16.95"	26.61"	33.9"
Conduit #3	66"	56.75"	20.95"	32.89"	41.9"
Conduit #4	70"	60.75"	24.95"	39.19"	49.9"

Use the same formula you used above to determine the length of the fourth conduit: stub + tail + developed length – 2 × radius = conduit length:

Conduit #4   70 + 60.75 + 39.17 – 49.9 = 120"

Conduit #3   66 + 56.75 + 32.19 – 41.9 = 113.74"

Conduit #2   62 + 52.75 + 26.61 – 33.9 = 107.46"

Conduit #1   58 + 48.75 + 20.33 – 25.9 = 101.18"

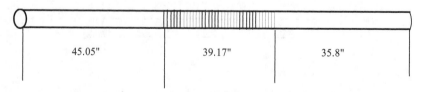

45.05"   39.17"   35.8"

Figure 10-18

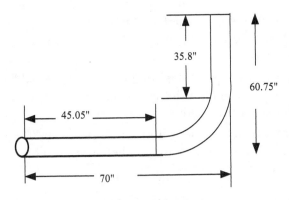

Figure 10-19

These conduits can be cut and prepared for bending. The fourth conduit is already bent. Take the third conduit and subtract the radius from the stub: 66 – 20.95 = 45.05″, the same as the fourth conduit. Measure from the end of the conduit to 45.05″ or 45¹⁄₁₆″. From this mark, measure out the developed length; for conduit #3 it will be 32.89″ or 32⁷⁄₈″. Mark off conduits #1 and #2 the same way. The starting point wil be 45¹⁄₁₆″ for all the conduits in this series, but the developed length changes. Conduit #2 = 45¹⁄₁₆ + 26⁵⁄₈; conduit #1 will be 45¹⁄₁₆ + 20⁵⁄₁₆″:

Conduit # 1	45¹⁄₁₆ + 20⁵⁄₁₆ = 65³⁄₈″
Conduit # 2	45¹⁄₁₆ + 26⁹⁄₁₆ = 71⁵⁄₈″
Conduit # 3	45¹⁄₁₆ + 32⁷⁄₈ = 77¹⁵⁄₁₆″
Conduit # 4	45¹⁄₁₆ + 39³⁄₁₆ = 84¼″

The cut conduits are as follows:

Conduit # 1	101³⁄₁₆ – 65³⁄₈ = 35¹³⁄₁₆″
Conduit # 2	107½ – 71⁵⁄₈ = 35⁷⁄₈″
Conduit # 3	113¾ – 77¹⁵⁄₁₆ = 35¹³⁄₁₆″
Conduit # 4	120 – 84¼ = 35¾″

Notice that the marks from the ends of the conduits are the same: 45¹⁄₁₆″ and 35¾″. There is a slight error in rounding off. Now conduits #1, #2, and #3 are ready to be laid out:

Conduit #1: Developed length of 20⁵/₁₆″. Divide 20⁵/₁₆″ by 18, which is 1⅛″. Mark the conduit using either surgical tubing or a ruler, as done with conduit #4.

Conduit #2: Developed length of 26⅝″. Divide 26⁵/₁₆″ by 30, which is ⅞″. Mark the conduit using either surgical tubing or a ruler, as done with the other two conduits.

Conduit #3: Developed length of 32⅞″. Divide 32⅞″ by 30, which is 1¹/₁₆″ Mark the conduit using either surgical tubing or a ruler, as done with the other three conduits.

Now you are ready to bend. Conduit #1 will have 18 5-degree bends. Conduits 2 and 3 will have 30 3-degree bends. When the bending is complete, lay the conduits on the floor with the proper spacing. The conduits will match up on both ends and the 4″ spacing will be maintained throughout. The results are well worth the effort.

Once you master segment bending, you can shorten the steps considerably. For example, three 2″ conduits are in a rack at 90 degrees. The first 90 (or inside 90) requires a 48″ stub. Spacing is 4″ center to center. The tail doesn't matter, as long as the conduits match on the ends.

First, determine the radius—8 × diameter of conduit + half of the outside diameter of conduit:

$$8 \times 2 + 1.25 = 17.25''$$

Now determine the radius and developed length of all 3 conduits:

Conduit	Radius	Radius × 1.57	= Developed Length
1	17.25	17.25 × 1.57	= 27.0825
2	17.25 + 4 = 21.25	21.25 × 1.57	= 33.36
3	17.25 + 4 = 25.25	25.25 × 1.57	= 39.64

To find the conduit length, multiply the spacing × 1.57 for the second conduit and 2 × spacing for the third conduit:

Second conduit 4 × 1.57 = 6.28″

Third conduit 8 × 1.57 = 12.56″

The first conduit is 12.56″ shorter than the third conduit, so take a full conduit (120″) and subtract 12.56″, or 120 − 12.56 = 107.44″. The second conduit will be 107.44 + 6.28 = 113.72″.

Now cut and thread the conduits. Once this is done, determine the starting point for the developed length:, stub – radius or 48 – 17.25 = 30.75 for the first conduit. This will be the starting point for all three conduits in this series.

Lay out the conduits as before. Measure out 30.75″ from the end of the first conduit. From this mark, measure off the developed length of the first conduit, which is 27.08″. Make your second mark here (see Figure 10-20). Divide 27.08″ by 30 = 0.90 inches. Mark the segments, then bend this conduit in 30 segments, 3 degrees each.

Second conduit. First measure and mark 30.75″ from the end of the conduit. From this mark, measure the developed length of 33.36″ (see Figure 10-21). Divide 33.36″ by 30 = 1.112″. Mark the segments and bend this conduit in 30 segments, 3 degrees each.

Third conduit. First measure and mark 30.75″ from the end of the conduit (see Figure 10-22). From this mark, measure the developed length of 39.64″. Divide 39.64″ by 30 = 1.32″ or 1⅝₁₆″. Mark the segments. Bend the conduit in 30 segments, 3 degrees each.

Notice from Figures 10-20 to 10-22 that the end measurements are constant and only the developed length expands by 6.28″ with each successive conduit. The 6.28″ expansion is nothing more than spacing × 1.57 or 4 × 1.57 = 6.28″.

It should be noted that often a group of conduits are of different sizes, but this doesn't matter. Establish the radius for the largest conduit in the group, then either add or subtract the spacing depending on where that conduit lays in the group. Smaller conduits can be less than the established radius; however, make sure that the radius never falls below 8 × diameter + ½ of the outside diameter of that conduit. It doesn't matter if different segments are used within a group. It is a good idea though to keep the segments between ¾ and 1¾″. To do this, you can use the seg-

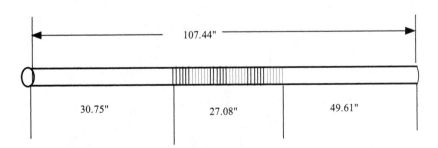

Figure 10-20

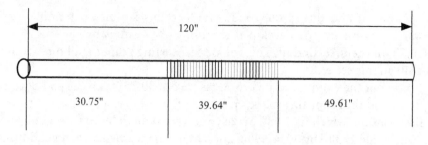

120"

30.75"   39.64"   49.61"

Figure 10-21

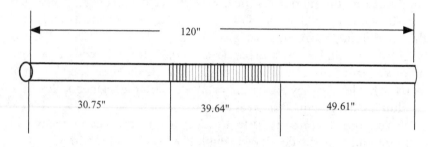

120"

30.75"   39.64"   49.61"

Figure 10-22

ment that works the best on a particular conduit within a group. With electronic benders, the job is a lot easier; however, a good eye and a protractor will give equally good results.

Occasionally, an existing conduit or process pipe must be followed. To accomplish this, measure the arc of the bend (the curved portion only; see Figure 10-23). This becomes the developed length of the bend. Another way to calculate the arc is to multiply a 90 degree arc by 4 to get the circumference of a full circle of 360 degrees. If the arc is 10 inches, the cir-

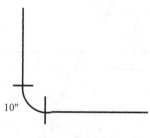

10"

Figure 10-23

cumference is 4 × 10 = 40 inches. To find the radius, use the formula, circumference = $2\pi r$, or $40/(2\pi) = r$, or 40/2 × 3.14 = 40/6.28 = 6.36 inches. Multiply the radius × 1.57 to get the developed length: 6.36 × 1.57 = 10. Divide the developed length by a multiple of 9, 18, 30, or 36 and mark the conduit for each segment. In the case of 9 marks, we will place our marks 1.11 inches apart. Each segment will be 10 degrees each and will yield a 90 degree bend with a 6.36 inch radius and a 10 inch arc. With this application, the conduit you bend will fit tightly to the pipe being followed. If a desired length for the stub is necessary, divide the arc of 10 inches by 1.57 to obtain the radius, which is subtracted from the stub. In other words, if a 36 inch stub is needed, subtract the radius, 6.36 inches, from 36 and mark the conduit: 36 − 6.36 = 29.64 inches. This will be the first mark of the bend. Now make 9 marks 1.11 inches apart away from the stub.

A back-to-back 90 is needed in the following problem and the conduit must fit snuggly to the pipe that is being followed. The first 90 is an inside radius and the second bend is an outside radius (see Figure 10-24).

First, change the arc to radius for both bends: Bend A has an arc of 10″. To find the radius, divide 10 by 1.57 = 6.36. Bend B has an arc of 15″. Divide 15 by 1.57 = 9.55″. To determine the length of the conduit with a 36″ stub and a 24″ tail, subtract the first radius from the desired stub, then add the developed length of the first bend of 10″:

$$36 - 6.36 + 10 = 39.64 \text{ inches}$$

Next subtract the first radius from the desired back-to-back distance of 36″:

$$36 - 6.36 = 29.64 \text{ inches}$$

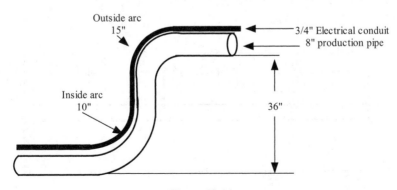

Figure 10-24

Next subtract the second radius from 29.72:

$$29.64 - 9.55 = 20.09 \text{ inches}$$

Add this to the previous length of 39.64 inches:

$$39.64 + 20.09 = 59.73 \text{ inches}$$

Next add to 59.73 inches the developed length of the second bend:

$$59.73 + 15 = 74.73 \text{ inches}$$

Since a 24 inch tail is needed, subtract the radius of the second bend from the desired length of the tail (24 inches):

$$24 - 9.55 = 14.45 \text{ inches}$$

Add the result to the previous length of the conduit:

$$74.73 + 14.45 = 89.18 \text{ inches}$$

To summarize (see Figure 10-25):

A   Stub = 36″. First radius: 36 − 6.36 = 29.64″
B   Developed length of first bend: 6.36 × 1.57 = 10″
C   Back-to-back distance: 36″ − first radius − second radius. 36 − 6.36 − 9.55 = 20.09
D   Developed length of second bend: 15″
E   Tail = 24″ − second radius: 24 − 9.55 = 14.45″

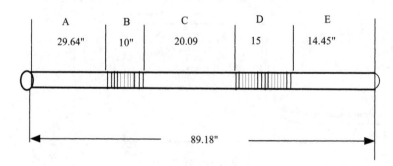

Figure 10-25

Add A + B + C + D + E = length of conduit before bending: 29.64 + 10 + 20.09 + 15 + 14.45 = 89.18″

Another way to find the length of the conduit before bending is to use $L + H + DL - 2r$ (where L = length, H = height, and DL = developed length). In this case, using the back-to-back 90s in the above example would yield:

$L + H$ + back-to-back distance + DL of 1st bend + (DL of 2nd bend– $2r$ of first bend) – $2r$ of 2nd bend = length of conduit before bending

or   36 + 24 + 36 + 10 + 15 – 12.72 – 19.10 = 89.18″

10-8   CIRCULAR BENDS ON LARGE VESSELS

Sometimes, it is necessary to bend a conduit around a large circular vessel. In order to do this, if the circumference or diameter of the vessel is not known, measure around the outside of the vessel to get the circumference. For example, a large tank measures 35 feet around, which is the circumference. Next, find the radius by using the formula, circumference = $2\pi r$:

$$35/(2 \times 3.14) = r \text{ or } 35/6.28 = 5.57 \text{ feet}$$

Convert 5.57 feet to inches = 5.57 × 12 = 66.87 inches. The radius equals 66.87 inches. If the conduit is 6 inches from the vessel, add 6 inches to the radius: 6 + 66.87 = 72.87 inches.

The circumference of the vessel, including 6 inches from the tank, is now 2 × 72.87 × 3.14 = 457.62 inches or 38.14 feet. Using the calculation in inches, 457.62 divided by 360 degrees is 1 degree every 1.27 inches. If a conduit is needed 18 feet around the tank, first find out how many degrees there are in the 18 foot arc of the conduit run. Convert the 18 foot arc to inches: 18 × 12 = 216 inches. Next, divide the arc by 1.27 inches per degree: 216/1.27 = 170 degrees. This will be the number of degrees in the bend. Next, figure how many degrees can be bent in a 120 inch conduit: 120/1.27 = 94.5 degrees. Next, to figure the distance between bends, select a multiple of 9, 18, 30, 36, 45, or 60 segments for a 90 degree bend. Divide 60 into the radius times 1.57 = 72.87 × 1.27 = 114.4 inches. Divide 114.4 inches by 60 to get the distance between segments: 114.4/60 = 1.9 inches between segments. Next divide 120 by 1.90

= 63 bends at 90/60 = 1.5 degrees per segment. The second conduit will be 75.5 degrees, still at 1.9 inches between segments. To find the length of the conduit needed for the second bend, either subtract 120 from 216 (216 − 120 = 96) or multiply radius × degrees of bend × 0.01745 = 96 inches. Next, divide 96 by 1.9 = 50. This is the number of segments in the second conduit.

Note: 0.01745 is 1.57/90. This can be used to find the length of the arc or, conversely, the degrees of the arc. The circumference of a circle is 360 degrees. Divide 6.28/360 = 0.01745 per degree.

Example. If an arc is 8 feet and the radius is 72.87 inches, what are the degrees of the arc? 8 × 12 = 96/0.01745 × 72.87 degrees.

How long is the arc if the radius is 72.87 at 30 degrees? 30 × 0.01745 × 72.87 = 38.14 inches.

On a Chicago bender, if the bending marks line up with the front of the shoe, the bend will not be where needed. The actual bending starts behind this point and will vary from bender to bender. This can easily be found with a practice bend. In order to bend at either end of the conduit (to get the bend as close to the end as possible), add a coupling and nipple to extend the conduit. If the shoe is too small to accommodate the coupling, bend the conduit with the next larger shoe.

CHAPTER 10 TEST

1. What is the shrinkage for a 25 degree offset 3 inches deep?
2. If a ½ inch rigid conduit has a 3½ inch gain and a 5 inch deduction on a 90 degree bend and, in addition, the conduit has a 15 inch stub 90 with a 30 degree offset with a 6 inch rise, what is the length of the conduit before bending if the overall length must be 55 inches?
3. Three ½ inch conduits are paralleled together, each has a 15 degree offset for a 3 inch offset and a spacing of 3 inches. The first offset on conduit A starts at 8 inches from the end of the conduit. In order for these conduits to match, what is the starting point of the offsets on conduit B and conduit C?
4. A 5 inch kick is needed 15 inches back from the 90. What is the angle of the kick?

5. A conduit must cross a 3-inch-high obstruction. From the center of the obstruction, how far is each of the outside bends on a three-point saddle?

6. A back-to-back 90 is needed. There is a 5 inch deduction for the 90 and the gain is 3½ inches. 36 inches are required between the 90s and each 90 is 15 inches. How long is the conduit before bending?

7. A 45 degree offset with a 10 inch rise is bent on a 60 inch piece of conduit. What is the length of the conduit after the bend?

8. What is the minimum radius of a 2 inch EMT conduit?

9. If a second conduit is paralleled with the 2 inch conduit in Problem 8, what would be the radius if 4 inch spacing is to be maintained?

10. What is the developed length of the conduit in Problem 9?

11. If the 2 inch conduit in Problem 8 had a stub of 36 inches and a tail of 48 inches, what is the length of the conduit before bending?

12. What is the length of the second conduit in Problem 9 with a stub of 36 inches and a tail of 48 inches?

13. A three-point saddle has three bends. What degrees are each of the bends?

14. In Problem 5, what is the shrinkage after the bends?

15. What is the cosecant of 25 degrees?

16. What is the cotangent of 25 degrees?

17. What is the shrinkage per inch of a 25 degree offset?

18. A right triangle is formed on any offset. The three sides of the triangle are the hypotenuse, opposite side, and adjacent side. If an offset is 30 degrees with a 6 inch rise, what are the hypotenuse and opposite side in inches?

19. What is the adjacent side in inches in Problem 18?

20. If shrinkage can be determined by subtracting the adjacent side from the hypotenuse, what is the shrinkage in Problem 18?

21. The circumference of a tank is 15 feet. If a conduit goes a quarter of the way around the tank at a distance of 6 inches from the tank, what is the circumference of the conduit run in inches?

22. How many inches are in each degree in Problem 21?

23. The conduit is going a quarter of the way around the tank in Problem 21. How many inches long will the conduit be?

24. If 2 degrees were used for each segment of the bend, what would be the distance between each segment?

25. A ½ inch conduit has three 90 degree bends. Assume that there is a 3½ inch gain for each 90. What would be the length of the conduit before bending if the stub were 18 inches, the tail 24 inches, and the middle 90 20 inches from the first 90 and 36 inches from the last 90?

MATH REVIEW

11-1 FRACTIONS

Fractions are important in electrical calculations. A good electrician must be able to add, subtract, multiply, and divide fractions easily. A fraction is divided into two parts: the denominator, the total number of parts of the fraction; and the numerator, the number of parts of the fraction that are divided by the denominator:

$$\frac{\text{Numerator}}{\text{Denominator}}$$

In the fraction 2/3, the total number of parts is 3. The number of parts of this fraction that are divided by the denominator is 2.

The fraction 2/3 is represented in Figure 11-1. The total number of parts in this fraction is the circle divided into 3 parts, which is the denominator. The shaded area of the circle represents the numerator, the 2 parts of this fraction that are divided by the denominator.

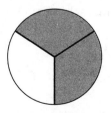

Figure 11-1

To add fractions together, a common denominator must be found. For example:

$$\frac{1}{2} + \frac{1}{4} = ?$$

In this example the smallest denominator, 2, will divide evenly into the second denominator, 4. If the numerator and the denominator are multiplied equally by the same number, the ratio of the fraction will remain intact:

$$\frac{1}{2} \times \frac{2}{2} = \frac{2}{4}$$

$\frac{2}{4}$ is equal to $\frac{1}{2}$, but now can easily be added to $\frac{1}{4}$:

$$\frac{2}{4} + \frac{1}{4} = \frac{3}{4}$$

When several fractions are added or subtracted together, a common denominator must be found before the operation can continue:

$$\frac{1}{3} + \frac{3}{4} + \frac{5}{8} - \frac{5}{6} = ?$$

The easiest way to find a common denominator is to multiply the denominators together: $3 \times 4 \times 8 \times 6 = 576$. This is becomes a cumbersome denominator. Another way would be to multiply the smallest denominator and the largest denominator together to see if all of the denominators would divide evenly into the new denominator: $3 \times 8 = 24$. 3, 4, 8, and 6 all go into 24 evenly. If the lowest denominator were 2, then the smallest denominator times the largest denominator would not work if 3 were one of the denominators. However, with a little practice the lowest common denominator can be found easily. Since 24 is the lowest common denominator for the above problem, each fraction must be converted to that denominator:

$$\frac{24}{3} = 8, \qquad \frac{24}{4} = 6, \qquad \frac{24}{8} = 3, \qquad \frac{24}{6} = 4$$

The numerator of each fraction must be multiplied by the same number that is multiplied by the denominator to equal the new denominator, 24:

$$\frac{1}{3} \times \frac{8}{8} = \frac{8}{24} \qquad \frac{3}{4} \times \frac{6}{6} = \frac{18}{24} \qquad \frac{5}{8} \times \frac{3}{3} = \frac{15}{24} \qquad \frac{5}{6} \times \frac{4}{4} = \frac{20}{24}$$

One way to check yourself to make sure that the fraction was correctly converted is to cross multiply. In the example of $\frac{1}{3}$ the equivalent fraction was $\frac{8}{24}$. To double check the fraction, multiply the numerator of the first fraction by the denominator of the second fraction, then multiply the other numerator of the second fraction by the denominator of the first fraction. The answer should be the same.

$$\frac{1}{3} \diagtimes \frac{8}{24}$$

$$1 \times 24 = 3 \times 8 \qquad \text{or} \qquad 24 = 24$$

Now that the fractions have been converted to fractions with the same common denominator, they can be easily be added together:

$$\frac{8}{24} + \frac{18}{24} + \frac{15}{24} - \frac{20}{24} = \frac{21}{24}$$

To reduce this fraction down to the lowest term, find a value that will divide evenly into both the numerator and the denominator:

$$\frac{21}{24} \div \frac{3}{3} = \frac{7}{8}$$

A mixed number is a whole number plus a fraction: $1\frac{3}{4}$.

An improper fraction is a fraction that has a larger numerator than its denominator: $\frac{7}{4}$.

If an improper fraction needs to be converted to a mixed number, find how many times the denominator will go into the numerator of the improper fraction. Four will go into 7 1 time with 3 remaining, so the mixed number becomes $1\frac{3}{4}$.

To convert a mixed number to an improper fraction, multiply the whole number by the denominator and add the numerator. For example, for $2\frac{7}{8}$,

$$\text{Multiply } 2 \times 8 = 16 + 7 = \frac{23}{8}$$

To multiply a fraction, multiply the numerator by the number and reduce to lowest terms:

$$\frac{5}{8} \times 5 = \frac{25}{8} = 3\frac{1}{8}$$

To divide a fraction, multiply the denominator by the number:

$$\frac{3}{4} \div 3 = \frac{3}{12} = \frac{1}{4}$$

Another way to divide is to change the divisor to its reciprocal and reduce, then multiply:

$$\frac{1\cancel{3}}{4} - \frac{1}{\cancel{3}_1} = \frac{1}{4}$$

The reciprocal of a number is the inverse of that number, and when the number and the inverse of that number are multiplied together, the product or answer is 1. The reciprocal of 2 is $\frac{1}{2}$:

$$\frac{2}{1} \times \frac{1}{2} = 1$$

The reciprocal of 3 is $\frac{1}{3}$ and the reciprocal of 4 is $\frac{1}{4}$. The reciprocal of $\frac{1}{5}$ is 5.

Exercise 11-1

1. Add $\frac{1}{4} + \frac{1}{3}$.

2. Add $\frac{1}{2} + \frac{1}{4} + \frac{1}{8}$.

3. What is the lowest term for the fraction $\frac{18}{24}$?

4. What is $7 \times \frac{3}{8}$? *Answer:* $\frac{3}{8} \times 7 = \frac{21}{8}$ or $2\frac{5}{8}$.

5. What is $\frac{1}{2}$ of $\frac{7}{8}$?

6. What is the reciprocal of 8?

7. Divide $\frac{5}{8}$ by 3.

8. Change $\frac{36}{24}$ to a mixed number.

9. A conduit will have 36 marks, each mark is $\frac{5}{8}$ inches apart. What is the distance from the beginning mark to the final mark?

10. Add $3\frac{9}{16} + 2\frac{3}{4}$.

Solutions to Exercise 11-1

1. The lowest common denominator is $3 \times 4 = 12$. $\frac{1}{4} = \frac{1}{4} \times \frac{3}{3} = \frac{3}{12}$; $\frac{1}{3} \times \frac{4}{4} = \frac{4}{12}$. $\frac{3}{12} + \frac{4}{12} = \frac{7}{12}$.

2. The lowest common denominator is 8. 2 goes into 8 4 times: $\frac{1}{2} \times \frac{4}{4} = \frac{4}{8}$. 4 goes into 8 2 times: $\frac{1}{4} \times \frac{2}{2} = \frac{2}{8}$. The new fractions can be added together: $\frac{4}{8} + \frac{2}{8} + \frac{1}{8} = \frac{7}{8}$.

3. 6 is the largest that will go into the denominator and the numerator evenly: $\frac{18}{24} \div \frac{6}{6} = \frac{3}{4}$.

4. $\frac{3}{8} \times 7 = \frac{21}{8}$ or $2\frac{5}{8}$.

5. $\frac{7}{8} \times \frac{1}{2} = \frac{7}{16}$.

6. The reciprocal of 8 is $\frac{1}{8}$.

7. $\frac{5}{8} \times \frac{1}{3} = \frac{5}{24}$.

8. Divide the numerator and denominator by 24. 24 will go into 36 1 time with a remainder of 12: $1\frac{12}{24}$. $\frac{12}{24}$ can be reduced down to $\frac{1}{2}$ by dividing 12 into the numerator and denominator; the final answer would be $1\frac{1}{2}$.

9. $\frac{5}{8} \times 36 = \frac{5}{8} \times \frac{36}{1} = \frac{180}{8} = 22.5$. The marks would be 22.5″ apart.

10. $3 + 2 = 5$ $\frac{3}{4} \times \frac{4}{4} = \frac{12}{16} + \frac{9}{16} = \frac{21}{16}$ or $1\frac{5}{16} + 5 = 6\frac{5}{16}$.

11-2 PERCENTAGE

A percentage is a part of a whole or another way of expressing a fraction. If a 20 ampere circuit is continuous duty, then the load on that circuit must be calculated at 125%. If the load is 15 amperes, it must be multiplied by 125%: 15 × 125% = 18.75 amperes.

The reciprocal of 125% is found by dividing 1.25 into 1: $125)\overline{100}^{\,0.80}$. 80% × 20 amperes = 16 amperes is the maximum continuous duty load that can be put on a 20 ampere breaker.

Percentage can be expressed in decimal form: 20% = 0.20 or 20 ÷ 100. 60% = 0.60. If a percentage is over 100%, then it will be expressed as a whole number and a decimal: 125% is 1.25.

To find 5% of 250, multiply 0.05 × 250 = 12.5. To add 5% to 150, multiply 150 × 1.05 = 157.5 or 150 × 0.05 = 7.5 + 150 = 157.5. Multiplying 150 by 1.05 saves a step in the process. The 1 in the 1.05, when multiplied out, gives the 150 and the 0.05 increases the number by 5%.

To subtract 5% from 150, multiply 150 × 0.95 = 142.5 or 150 × 0.05 = 7.50; 150 − 7.5 = 142.5. By subtracting 5% from 100% and subtracting the answer from 150, a step is eliminated.

To change a fraction to a percent, divide the denominator into 100, then multiply the numerator by the answer and divide by 100: $\frac{3}{4} = 4)\overline{100}^{\,25}$ 25 × 3 = 75%. An easier way to find the answer is to divide the numerator by the denominator and multiply by 100: $\frac{3}{4} = 4)\overline{3.00}^{\,0.75}$ 0.75 × 100 = 75%.

A number greater than 100% is like a mixed number: 135% = $1\frac{35}{100}$. If the conductors must be 135% of the load, as in the case of conductors feeding a capacitor, then they would be rated at the load plus 35%. If the load were 50 amperes, the conductors would have to be capable of carrying 67.5 amperes: 1.35 × 50 = 67.5.

To turn a number into a percent, divide the number by 100: $100)\overline{7.00}^{\,0.07}$; or invert 100 and multiply: $7 \times \frac{1}{100} = \frac{7}{100} = 100)\overline{7.00}^{\,0.07}$ 135% in decimal form would be 135 ÷ 100 = 1.35 or $135 \times \frac{1}{100} = 1.35$.

Some fractions converted to percent follow:

$\frac{1}{3} = 3)\overline{1.00}^{\,0.33}$ $\frac{1}{8} = 8)\overline{1.000}^{\,0.125}$ $\frac{2}{8}$ or $\frac{1}{4} = 8)\overline{2.00}^{\,0.25}$

$\frac{4}{8}$ or $\frac{2}{4}$ or $\frac{1}{2} = 8)\overline{4.00}^{\,0.50}$ $\frac{6}{8}$ or $\frac{3}{4} = 8)\overline{6.00}^{\,0.75}$

$$\frac{2}{3} = 3\overline{)2.00}^{\,0.66} \qquad \frac{1}{4} = 4\overline{)1.00}^{\,0.25} \qquad \frac{3}{8} = 8\overline{)3.000}^{\,0.375}$$

$$\frac{5}{8} = 8\overline{)5.000}^{\,0.625} \qquad \frac{7}{8} = 8\overline{)7.000}^{\,0.875}$$

To convert a mixed number to a percentage, first change the mixed number to an improper fraction, then divide the numerator by the denominator: $2\frac{3}{8} = 2 \times 8 = \frac{16}{8} + \frac{3}{8} = \frac{19}{8}; \frac{19}{8} = 8\overline{)19.000}^{\,2.375}; 2.375 \times 100 = 237.5\%.$

Exercise 11-2

1. Convert 36% to a decimal
2. Convert 1.25 into a percent.
3. What is the percentage of $\frac{7}{8}$?
4. Add 15% to 200.
5. What is the reciprocal of 0.25?
6. Add 6% to 180.
7. 125% of 60 is what?
8. Show $\frac{1}{2}$ of 1% in decimal form.
9. Subtract 15% from 90.
10. What is 25% of 75?

Solutions to Exercise 11-2

1. $100\overline{)36.00}^{\,0.36} = 0.36$
2. $1.25 \times 100 = 125\%$
3. $8\overline{)7.000}^{\,0.875} = 0.875 \times 100 = 87.5\%$
4. $200 \times 1.15 = 230$
5. $1 \div 0.25 = 4$
6. $180 \times 1.06 = 190.8$ or $180 \times 0.06 = 10.8 + 180 = 190.8$

7. $100\overline{)125.00}^{1.25} = 1.25 \times 60 = 75$

8. $\dfrac{1}{2} \times \dfrac{1}{100} = \dfrac{1}{200} \; 200\overline{)1.000}^{0.005} = 0.005$

9. $90 \times 0.85 = 76.5$ or $90 \times 0.15 = 13.5$; $90 - 13.5 = 76.5$

10. $75 \times 0.25 = 18.75$

11-3 DECIMALS

Decimals are another form of fractions. The first number to the right of the decimal point is tenths. The second number to the right of the decimal point is hundredths. The third number to the right of the decimal is thousandths. The fourth number to the right of the decimal point is ten thousandths and so on:

$$(1)\; 0.3 = \frac{3}{10} \qquad (2)\; 0.33 = \frac{33}{100} \qquad (3)\; 0.333 = \frac{333}{1000} \qquad (4)\; 0.3333 = \frac{3333}{10,000}$$

The denominators in the above examples are in multiples of 10:

$$(1)\; 10 \qquad (2)\; 10 \times 10 = 100 \qquad (3)\; 10 \times 10 \times 10 = 1000$$

$$(4)\; 10 \times 10 \times 10 \times 10 = 10,000$$

To add decimals together, the decimal points must line up and the tenths must line up with the tenths, the hundredths with the hundredths, and so on. It can make it easier if zeros are added to the right of the decimal point to help keep the numbers lined up correctly. Remember, adding zeros to the right of the decimal does not change the number. To add 0.3 to 0.33:

$$\begin{array}{r} 0.33 \\ +0.30 \\ \hline 0.63 \end{array}$$

Adding a zero to the right of the decimal adds balance:

$$\begin{array}{r} 0.33 \\ +0.30 \\ \hline 0.63 \end{array}$$

Subtract decimals from each other is performed similarly:

$$0.33 - 0.3 = \quad \begin{array}{r} 0.33 \\ -\,0.3 \\ \hline 0.03 \end{array} \qquad \text{or} \qquad \begin{array}{r} 0.33 \\ -\,0.30 \\ \hline 0.03 \end{array}$$

After practice, it will become second nature to subtract directly by just adding the zero to 0.3. Remember 0.3 and 0.30 are the same: $0.3 = \dfrac{3}{10}$; $0.30 = \dfrac{30}{100}$. $\dfrac{30}{100}$ reduced to the lowest terms is $\dfrac{3}{10}$.

To multiply decimals together, multiply the numbers just like whole numbers:

$$\begin{array}{r} 5.5 \\ \times\,3.6 \\ \hline 1980 \end{array}$$

Add the numbers to the right of the decimals in each number. There is one in each, two total. The decimal must be placed two places to the left of the last number. The correct placement of the decimal would be between the 9 and the 8: 19.80
↑

Another example is

$$\begin{array}{r} 12.25 \\ \times\,6.45 \\ \hline 790125 \end{array}$$

Since there are two numbers to the right of the decimal in each of the numbers to be multiplied together, they are added together: $2 + 2 = 4$. The decimal point will go four places from the left of the last number, between the 9 and the 0: 79.0125.
↑

To divide decimals by whole numbers, the decimal point goes directly above the dividend (the number being divided) into the quotient (the answer):

$$\overset{\text{Quotient}}{\text{Divisor}\,\overline{)\text{Divident}}}$$

Example:

$$5\,\overline{)1.25}^{\,0.25}$$

To divide a decimal by another decimal, the decimal must be removed from the divisor to make it a whole number. What is done to the divisor must also be done to the dividend: $0.16 \div 0.08 = \dfrac{0.16}{0.08}$. To remove the decimal from the denominator, multiply the numerator and the denominator by the same number, in this case, 100:

$$\frac{0.16}{0.08} \times \frac{100}{100} = \frac{16}{8} = 2 \qquad \text{or} \qquad 0.8\overline{)0.16} = 8\overline{)16}\,^{2} = 2$$

Exercise 11-3

1. Add 0.63, 0.5, 1.25, and 0.05.
2. Divide 5 into 65.75.
3. Divide 0.05 into 0.75
4. Multiply 0.60 × 0.40.
5. Change 40% to a decimal.
6. Subtract 0.65 from 0.90.
7. Change $\dfrac{5}{8}$ to a decimal.
8. Change $\dfrac{3}{4}$ to a decimal.
9. Subtract 0.60 from 1.
10. Zeros to the right of the decimal point, as in 0.2400, change the value. True or False?

Solutions to Exercise 11-3

1.   0.63
      0.50
      1.25
   + 0.05
      2.43

2. $5\overline{)65.75}\,^{13.15}$

3. $0.05\overline{)0.75} = 5\overline{)75}; \; 5\overline{)75}\,^{15}.$

4. $0.60 \times 0.40 = 0.2400$ or 0.24

5.
$$\begin{array}{r} 0.40 \\ 100\overline{)40.00} \end{array}$$

6.
$$\begin{array}{r} 0.90 \\ -0.65 \\ \hline 0.25 \end{array}$$

7.
$$\begin{array}{r} 0.625 \\ 8\overline{)5.000} \end{array}$$

8.
$$\begin{array}{r} 0.75 \\ 4\overline{)3.00} \end{array}$$

9.
$$\begin{array}{r} 1.00 \\ -0.60 \\ \hline 0.40 \end{array}$$

10. False

11-4   BASIC ALGEBRA

Algebra uses equations made up of letters and/or numbers to derive general formulas that can solve specific problems. One common example of an algebraic equation is the one used to find the square footage of a room: $A = LW$, where L = length, W = width, and A = area. If the length is 15 feet and the width is 10 feet, $A = 15 \times 10$ or $A = 150 \text{ ft}^2$. The formula or equation is general because it can be applied to any rectangle. The equation becomes specific when values are assigned to the letters: $L = 15$ feet, $W = 10$ feet. $A = LW$ is the same as L × W. LW is the product of the two factors, L and W. If the variables (L and W) change, then A will change as well.

An important step in algebra is to transpose a formula. Transposition makes it possible to move variables and constants to solve for different values. Consider $X - 3 = 7$. To solve for X, the 3 must be moved to the other side of the equal sign. What is done on one side of the equal sign must be done on the other side of the equal sign. This keeps the equation proportional or in balance.

To move the −3 in the $X - 3$, add a +3:

$$X - 3 + 3 = X + 0 \qquad \text{or} \qquad X = X$$

Notice how the +3 and −3 cancel each other out. Since 3 was added to one side of the equal sign, +3 must be added to the other side of the equal sign: $7 + 3 = 10$.

The revised equation is now $X = 10$. Now X has been solved. To check the problem, substitute 10 for X in the original equation and both sides should be equal: $10 - 3 = 7$.

If the equation is $X + 4 = 6$, to isolate X, the 4 must be cancelled. To cancel a positive 4, -4 must be added:

$$X + 4 - 4 = X + 0 \text{ or } X$$

If 4 is subtracted from one side of the equal sign, 4 must be subtracted from the other side of the equal sign:

$$X + 4 - 4 = 6 - 4, \qquad X + 0 = 6 - 4 \text{ or } X = 2$$

To check the problem, substitute 2 for X in the original equation and both sides should be equal: $2 + 4 = 6$.

If the equation is $2X = 20$, the 2 must be cancelled. $2X$ is the product of the factors 2 and X or $2 \times X$. To cancel one of the factors, divide by the factor to be cancelled:

$$\frac{^1 2X}{^1 2} = \frac{1X}{1} \qquad \text{or } X$$

If the factor is cancelled by dividing on one side of the equal sign, then it must be divided by the same factor on the other side of the equal sign:

$$\frac{^1 2X}{_1 2} = \frac{20^{10}}{2_1} \qquad \text{or} \qquad \frac{1X}{1} = \frac{10}{1} \qquad \text{or} \qquad X = 10$$

To check the problem, substitute 10 for X in the original equation; then both sides should be equal: $2 \times 10 = 20$.

If the equation is $\frac{X}{3} = 7$, the 3 must be cancelled. Since 3 is a divisor, to cancel it out, multiplication is required:

$$\frac{^1 3}{1} \times \frac{X}{3_1} \qquad \text{or } X$$

Since the factor was cancelled by dividing on one side of the equal sign, then it must be divided by the same factor on the other side of the equal sign: $7 \times 3 = 21$.

Now the equation looks like this:

$$\frac{^1 3X}{3_1} = 7 \times 3 \qquad \text{or} \qquad X = 21$$

To check the equation, substitute 10 for X in the original equation:

$$\frac{21}{3} = 7$$

Review

Addition: $X - 7 = 3$
             $X - 7 + 7 = 3 + 7$
             $X + 0 = 10$
             $X = 10$

Subtraction: $X + 4 = 6$
             $X + 4 - 4 = 6 - 4$
             $X + 0 = 2$
             $X = 2$

Division: $2X = 20$. To cancel 2, divide by 2:

$$\frac{^{1}2X}{2_1} = \frac{20^{10}}{2_1} \quad \text{or} \quad X = 10$$

Multiplication: $\frac{X}{3} = 7$. To cancel 3, multiply by 3:

$$\frac{^{1}3}{1} \times \frac{X}{3_1} = 7 \times 3$$

$$X = 21$$

An example of algebra in electrical calculations is Ohm's Law. The equation $E = RI$ $(R \times I)$ is a good example. E = voltage, R = resistance, and I = amperage. The basic formula can be applied to several circuits. Even though the letters stand for variables, the basic rules still apply. A letter can be canceled the same way a number is. If a circuit has a total amperage of 5 amperes and a total resistance of 24 ohms, by substituting the numbers, E can be solved: $E = RI$ or $E = 24 \times 5 = E = 120$ volts. If the variables change, so does the answer, but the equation or formula stays the same. If $R = 12$ ohms and $I = 5$ amperes, what is the voltage? $E = RI$ still applies. Substitute the letters with the values and solve the equation: $E = 12 \times 5$ or $E = 60$ volts.

The advantage of the equation is that it can be used to solve for different values, not just voltage. Whatever numbers (values) are known, the missing value can be solved for using $E = RI$. If the voltage is known and the amperage is known, the equation can be reworked to find the resistance.

To find the resistance, R must be isolated on one side of the equal sign with E and I on the other side of the equal sign, like the previous example involving X. This is called transposition. $E = RI$. To isolate R on one side of the equal sign, I must be removed and put on the other side of the equal sign. To do that, I must be divided into RI and when I is divided on one side of the equal sign, it must be divided on the other side of the equal sign. This is called transposition:

$$\frac{E}{I} = \frac{R \times I^1}{I_1} = \frac{E}{I} = R$$

When I is divided into RI, the Is cancel out: $\frac{I}{I} = 1$. This is exactly the same as dividing any number by itself: $\frac{2}{2} = 1$.

Now the equation has been transposed to solve for R instead of E. If $E = 60$ and $I = 5$, what is the resistance? $\frac{E}{I} = R = \frac{60}{5}$ or $R = 12$.

Some equations require multiple steps to transpose an equation. The voltage drop equation is a good example: $VD = \frac{2KIL}{Cm}$. When $VD = 6$ volts and the amperage $= 22$ amperes, how long can a two-wire circuit made up of #10 copper wire go within the 6 volt drop limitation? $K = 11$ Cm $= 10,380$ (see Chapter 9, Table 8, Conductor Properties, *National Electrical Code®*). The formula must be transposed to isolate L so that the distance can be solved for:

$$VD = \frac{2KIL}{Cm} = VD \times Cm = \frac{2KIL}{Cm_1} \times \frac{Cm_1}{1} = VD \times Cm = 2KIL$$

Now L can be isolated by dividing $2KIL$ by $2KI$:

$$\frac{VD \times Cm}{2KI} = \frac{^1 2KIL}{2KI_1}$$

The new equation reads $\frac{VD \times Cm}{2KI} = L$. Now substitute the known values, $VD = 6$, Cm $= 10,380$, $K = 11$ and $I = 22$:

$$\frac{6 \times 10,380}{2 \times 11 \times 22} = L \quad \text{or} \quad \frac{62,280}{484} = L \text{ or } 128.67 \text{ feet}$$

To check the answer, use the values in the original equation:

$$VD = \frac{2KIL}{Cm} = 6 = \frac{2 \times 11 \times 22 \times 128.67}{10,380}$$

Being able to transpose an equation means that learning one equation can be used to solve five variables:

a) $VD = \dfrac{2KIL}{Cm}$ b) $Cm = \dfrac{2KIL}{VD}$ c) $L = \dfrac{VD \times Cm}{2KI}$

d) $I = \dfrac{VD \times Cm}{2KL}$ e) $K = \dfrac{VD \times Cm}{2IL}$

In the equations above, (b) through (e) were transposed from the first equation (a).

Exercise 11-4

1. Solve for X when $2X + 4 = 12$.

2. Solve for X when $\dfrac{6X}{3} = 10$.

3. Find the total resistance of a circuit when the voltage = 30 volts and amperage = 6 amperes.

4. Transposing a number or letter from one side of the equal sign must be done on the other side of the equal sign. True or False?

5. Solve for X when $\dfrac{X}{7} = 2$.

6. What is the reciprocal of X?

7. If an area of a room is 50 feet² and the length is two times the width, what is the length and width of the room. Use $A = LW$.

8. What size copper wire is needed for a 16 ampere load 250 feet from the source to maintain a 5% voltage drop at 120 volts (6 volts). Solve for Cm (Circular mills).

9. Check the following equation to see if $X = 12$:

$$\frac{X}{2} - 3 = 5$$

10. What are the factors of the product RI?

Solutions to Exercise 11-4

1. $2X + 4 - 4 = 12 - 4$; $\quad \dfrac{\cancel{2}X}{\cancel{2}1} = \dfrac{\cancel{8}^4}{\cancel{2}_1}$ $\quad$ or $X = 4$

2. $\dfrac{\cancel{6}^2 X}{\cancel{3}_1} = 10$; $\quad \dfrac{\cancel{2}^1 X}{\cancel{2}_1} = \dfrac{\cancel{10}^5}{\cancel{2}_1} = 5$; $\quad X = 5$

3. $E = RI$; $\quad \dfrac{E}{I} = \dfrac{R \times I}{I}$ $\quad$ or $\quad \dfrac{E}{I} = R \dfrac{30}{6} = 5$ ohms

4. True

5. $\dfrac{^1\cancel{7}}{1} \times \dfrac{X}{\cancel{7}_1} = 2 \times 7$ or $X = 14$

6. $\dfrac{1}{X}$

7. $A = 50$ feet^2, $L = 2W$, $W = W$, $A = 2W \times W$, or $50 = 2W^2$

$$\dfrac{^1\cancel{2}W^2}{\cancel{2}_1} = \dfrac{^{25}\cancel{50}}{\cancel{2}_1} = W^2 = 25$$

$$W^2 = W \times W = 25, \sqrt{W^2} = \sqrt{25} \; W = 5, L = 2W = 2 \times 5 = 10$$

8. $VD = \dfrac{2KIL}{Cm}$, $Cm \times VD = \dfrac{2KIL}{^1\cancel{Cm}} \times \dfrac{\cancel{Cm}^1}{1}$ or $\dfrac{Cm \times {}^1\cancel{VD}}{^1\cancel{VD}} = \dfrac{2KIL}{1} \times \dfrac{1}{VD}$

or $Cm = \dfrac{2KIL}{VD}$

$$Cm = \dfrac{2 \times 11 \times 16 \times 250}{6} = 14666.7 \; Cm$$

The wire size is larger than #10 (10,380 Cm) but smaller than #8 (16,510 Cm). See Chapter 9 Table 8, Conductor Properties, of the *National Electrical Code*®. #8 is correct.

9. $\dfrac{X}{2} - 3 = 5$. Substitute 12 for X. $\dfrac{12}{2} - 3 = 5$ or $\dfrac{12}{2} - 3 = 5$ or $\dfrac{^6\cancel{12}}{\cancel{2}_1} - 3 = 5$

or $3 = 5$. 3 does not $= 5$, so X does not $= 12$. $\dfrac{X}{2} - 3 = 5$ or $\dfrac{X}{2} - 3 + 3 = 5 + 3$ or $\dfrac{^1\cancel{2}}{1} \times \dfrac{X}{\cancel{2}_1} = 8 \times 2 = 16$. The correct answer is $X = 16$.

10. *RI* is the product of the factors *R* and *I* or $R \times I$.

11-5   ANGLES

Triangles and angles are important in electrical calculations. Angles are used in a wide variety of calculations. A triangle has three sides and three angles. The total of the angles in a triangle is 180 degrees. There are different types of angles:

Acute angle: more than 0 degrees but less than 90 degrees

Right angle: 90 degree angle

Obtuse angle: more than 90 degrees but less than 180 degrees

There are several types of triangles:

Isosceles triangle: Two sides have the same length and the angles opposite the equal sides are equal

Scalene triangle: No two sides are equal in length and no two angles are equal

Acute triangle: All three angles are less than 90 degrees

Obtuse triangle: one angle is greater than 90 degrees

Right triangle: one angle is a right angle or 90 degrees

The right triangle is the most useful to electricians (see Figure 11-2). The longest side is the hypotenuse or Side C.

Trigonometry is the mathematics of the sides and angles of triangles. The most important equation of a right triangle is the Pythagorean Theorem or Pythagoras's Theorem. The Theorem is $A^2 + B^2 = C^2$. Each letter in the equation represents a side of the triangle. Side C is the hypotenuse or the side opposite the 90 degree angle. Side A is the adjacent side. The third side is the opposite side, B. By knowing two sides, the third side can be found.

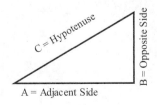

Figure 11-2

One application of the triangle is used in conduit bending. When bending a 30 degree offset with a 6 inch rise, the distance between marks is two times the rise or 12 inches (see Figure 11-3).

To find the third side, the equation $A^2 + B^2 = C^2$ is used. C is the hypotenuse and B is the opposite side. To solve for A, the equation must be transposed:

$$A^2 + \cancel{B^2} - \cancel{B^2} = C^2 - B^2 \qquad \text{or} \qquad A^2 = C^2 - B^2$$

$C = 12$ and $B = 6$. $A^2 = 144 - 36$ or $A^2 = 108$. $A = \sqrt{108}$ or 10.39 inches.

Angles can be determined by knowing the lengths of all the sides. There are six trigonometric functions to accomplish this:

1. $\dfrac{\text{opposite side}}{\text{hypotenuse}} = \emptyset \text{ sine}$
2. $\dfrac{\text{adjacent side}}{\text{hypotenuse}} = \emptyset \text{ cosine}$

3. $\dfrac{\text{opposite side}}{\text{adjacent side}} = \emptyset \text{ tangent}$
4. $\dfrac{\text{hypotenuse}}{\text{opposite side}} = \emptyset \text{ secant}$

5. $\dfrac{\text{hypotenuse}}{\text{adjacent side}} = \emptyset \text{ cosecant}$
6. $\dfrac{\text{adjacent side}}{\text{opposite side}} = \emptyset \text{ cotangent}$

These equations will not yield the angle directly. The result, however, can be taken to a chart of natural trigonometric functions (Table 11-1) and converted to the correct answer:

To find the sine: $\dfrac{\text{opposite side}}{\text{hypotenuse}} = 0 \text{ sine} = \dfrac{6}{12} = 0.5$

To find the cosine: $\dfrac{\text{adjacent side}}{\text{hypotenuse}} = 0 \text{ cosine} = \dfrac{10.39}{12} = 0.8660$

To find the tangent: $\dfrac{\text{opposite side}}{\text{adjacent side}} = 0 \text{ tangent} = \dfrac{6}{10.39} = 0.5774$

To find the secant: $\dfrac{\text{hypotenuse}}{\text{adjacent side}} = 0 \text{ secant} = \dfrac{12}{10.39} = 1.154$

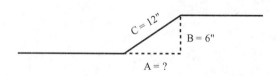

Figure 11-3

Table 11-1   Natural trigonometric functions

Angle	Sine	Cosine	Tangent	Cotangent	Secant	Cosecant
0	0	1	0		1	
1	0.0175	0.9998	0.0175	57.29	1.002	57.2987
2	0.0349	0.9994	0.0349	28.636	1.006	28.6537
3	0.0523	0.9986	0.0524	19.081	1.0014	19.107
4	0.0698	0.9976	0.0699	14.301	1.0024	14.336
5	0.0872	0.9962	0.0875	11.43	1.0038	11.474
6	0.1045	0.9945	0.1051	9.5144	1.0055	9.567
7	0.1219	0.9925	0.1228	8.1443	1.0075	8.202
8	0.1392	0.9903	0.1405	7.1154	1.0098	7.185
9	0.1564	0.9877	0.1584	6.3138	1.0125	6.392
10	0.1736	0.9848	0.1763	5.6713	1.0154	5.759
11	0.1908	0.9816	0.1944	5.1446	1.0187	5.241
12	0.2079	0.9781	0.2126	4.7046	1.0223	4.81
13	0.225	0.9744	0.2309	4.3315	1.0263	4.5
14	0.2419	0.9703	0.2493	4.0108	1.0306	4.134
15	0.2588	0.9659	0.2679	3.7321	1.0353	3.864
16	0.2756	0.9613	0.2867	3.4874	1.0403	3.628
17	0.2924	0.9563	0.3057	3.2709	1.0457	3.42
18	0.309	0.9511	0.3249	3.0777	1.0515	3.236
19	0.3256	0.9455	0.3443	2.9042	1.0576	3.072
20	0.342	0.9397	0.364	2.7475	1.0642	2.924
21	0.3584	0.9336	0.3839	2.6051	1.0711	2.79
22	0.3746	0.9272	0.404	2.4751	1.0785	2.674
23	0.3907	0.9205	0.4245	2.3559	1.0864	2.554
24	0.4067	0.9135	0.4452	2.246	1.0946	2.459
25	0.4226	0.9063	0.4663	2.1445	1.1034	2.366
26	0.4384	0.8988	0.4877	2.0503	1.1126	2.281
27	0.454	0.891	0.5095	1.9626	1.1223	2.203
28	0.4695	0.8829	0.5317	1.8807	1.1326	2.13
29	0.4848	0.8746	0.5543	1.804	1.1434	2.063
30	0.5	0.866	0.5774	1.7321	1.1547	2
31	0.515	0.8572	0.6009	1.6643	1.6666	1.942
32	0.5299	0.848	0.6249	1.6003	1.1792	1.887
33	0.5446	0.8387	0.6494	1.5399	1.1924	1.836
34	0.5592	0.829	0.6745	1.4826	1.2062	1.788
35	0.5736	0.8192	0.7002	1.4281	1.2208	1.743
36	0.5878	0.809	0.7265	1.3764	1.2361	1.701
37	0.6018	0.7986	0.7536	1.327	1.2521	1.662
38	0.6157	0.788	0.7813	1.2799	1.269	1.624
39	0.6293	0.7771	0.8098	1.2349	1.2868	1.589
40	0.6428	0.766	0.8391	1.1918	1.3054	1.556
41	0.6561	0.7547	0.8693	1.1504	1.325	1.527
42	0.6691	0.7431	0.9004	1.1108	1.3456	1.494
43	0.682	0.7314	0.9325	1.0724	1.3673	1.466
44	0.6947	0.7193	0.9657	1.0355	1.3902	1.44
45	0.7071	0.7071	1	1	1.4142	1.414

To find the cosecant: $\dfrac{\text{hypotenuse}}{\text{opposite side}} = 0$ cosecant $= \dfrac{12}{6} = 2$

The sine of an angle is the reciprocal of the cosecant: $0.5 \dfrac{1}{0.5} = 2$

The cosine of an angle is the reciprocal of the secant: $0.8660 = \dfrac{1}{0.8660}$ $= 1.154$

The tangent of an angle is the reciprocal of the cotangent $0.5774 = \dfrac{1}{0.5774} = 1.731$

If an offset is needed and the angle can't be determined but the rise is 3 inches and the distance between bends 9.71 inches, what is the degree of the offset? Rise = opposite side, distance between bends = hypotenuse. Rise = 3 inches, hypotenuse = 9.71 inches.

$$\frac{\text{hypotenuse}}{\text{opposite side}} = 0 \text{ cosecant} = \frac{9.71}{3} = 3.236$$

In the Table 11-1, look under the column Cosecant and find 3.236. Look directly across to Angle and read the angle: cosecant 3.236 = 18 degrees.

To find the multiplier for a 30 degree bend, press 30 on your calculator. Next press the sin button. The display will read 0.5. Next press $\dfrac{1}{X}$. The display will change to 2. This is the multiplier. Now press the multiplying key, then the height of your offset, in this case 6. The answer will be 12:

$$\boxed{30} \; + \; \boxed{\sin} \; + \; \boxed{\dfrac{1}{X}} \; + \; \boxed{*} \; + \; \boxed{6} \; = \; \boxed{12}$$

This is the distance between the bends. Put 12 in the plus memory. Now put 30 in the calculator display again. Press tan. The display will change to 0.5773. Now press $\dfrac{1}{X}$ and the display will change to 1.732. Press the multiplier key and then enter the height of the offset, 6. The display will change to 10.39:

$$\boxed{\tan} \; + \; \boxed{\dfrac{1}{X}} \; + \; \boxed{*} \; + \; \boxed{6} \; = \; \boxed{10.39}$$

If the sine, tangent, or cosecant of an angle is known, the degrees can be calculated by using a scientific calculator. To find an angle in degrees on the calculator, first put in the sine value of the angle: 0.5. Then press 2nd function, then press sin:

$$\boxed{.5} \quad + \quad \boxed{\begin{array}{c}\text{2nd}\\\text{func}\end{array}} \quad + \quad \boxed{\sin} \quad = \quad \boxed{30}$$

This will work for all of the trigonometric functions. Try dividing

$$\frac{\text{opposite side}}{\text{adjacent side}} = 0 \text{ tangen} \frac{6}{10.39} = 0.5774$$

Now enter 0.5774 in the calculator:

$$\boxed{.5774} \quad + \quad \boxed{\begin{array}{c}\text{2nd}\\\text{func}\end{array}} \quad + \quad \boxed{\tan} \quad = \quad \boxed{30}$$

With practice, this method will save time and will eliminate the need to carry the table of natural trigonometric functions around.

Exercise 11- 5

1. What are the three sides of a right triangle?
2. If a triangle has a 90 degree angle and a 30 degree angle, what is the remaining angle?
3. What is the reciprocal of the sine of an angle?
4. What is an equilateral triangle?
5. If the cosecant of an angle is 3.8637 what is the angle?
6. If the rise of an offset is 9 inches and the distance between bends is 12.726 inches, what is the angle of the offset?
7. When bending conduit, the marks between the bends are equivalent to what part of the triangle?
8. If the height of an offset is 5 inches and the distance between bends is 10 inches, what is the loss or shrinkage of the conduit?
9. What is the cosecant of a 25 degree angle?
10. If the hypotenuse of a triangle is 15 inches and the rise is 6 inches, what is the angle?

Solutions to Exercise 11-5

1. Hypotenuse, opposite side, and adjacent side
2. Maximum degrees in a triangle is 180. If two angles are 90 and 30 degrees (= total 120 degrees), then $180 - 120 = 60$ degrees.
3. Cosecant
4. An equilateral triangle has three equal sides and three equal angles.
5. 15 degrees (from the table of natural trigonometric functions)
6. $\dfrac{\text{hypotenuse}}{\text{opposite side}} = 0$ cosecant $\dfrac{9}{12.726} = 0.7072 = 45$ degrees (from the table of natural trigonometric functions)
7. Hypotenuse
8. $A^2 + B^2 = C^2, C = 10, B = 5$ $A^2 + B^2 - B^2 = C^2 - B^2, 100 - 25 = A^2, 75 = \sqrt{A^2}$ or 8.66 inches, 10 inches $-$ 8.66 inches $= 1.34$ inches.
9. 2.3662 (from the table of natural trigonometric functions)
10. $\dfrac{\text{hypotenuse}}{\text{opposite side}} = 0$ cosecant $\dfrac{15}{6} = 2.5 =$ approximately 23 degrees (from the table of natural trigonometric functions)

11-6   SQUARE ROOTS

The square of a number is the number times itself. The factors in a square of a number are $X \times X$ or X^2. The factor that makes a number a square is called the square root. The easiest way to find a square root of a number is to enter that number in a calculator and press the square root key. If a calculator isn't available, then the manual method will have to do.

To do square roots manually, divide the number to be squared into groups of 2. If there is an odd number, the first number is grouped by itself and the remainder of the numbers are grouped in pairs. For every decimal place in the answer, add two zeroes to the group. Example: find the square root of 3306.25 or $\sqrt{3306.25}$:

$$\sqrt{3306.25} \quad 33 \quad 06 \quad 25$$

Find a number times itself that will not be greater than 33:

$$5 \times 5 = 25$$

$$\begin{array}{r} 5 \cdot \\ \sqrt{3306.25} \\ \underline{25} \\ 10?\overline{)806} \end{array}$$

Bring the next group down, 06, to make the new number 806. Excluding the 6, how many times will 10 go into 80? 8 times. Add the 8 at the end of the 10, creating the new number of 108. Multiply $108 \times 8 = 864$. 864 is larger than 806 and will not work. Try 7 instead: $107 \times 7 = 749$. Since this number is less than 806, it is correct.

$$\begin{array}{r} 5\ 7.\ 5 \\ \sqrt{3306.25} \\ \underline{25} \\ 10?\overline{)806} \\ \\ \underline{749} \\ 114?\overline{)5725} \\ \\ \underline{5725} \\ 0 \end{array}$$

The partial answer is doubled and brought down as part of the new divisor: $57 \times 2 = 114$. How many times will 114 go into 572? Approximately 5 times. Change the ? to 5 and multiply $5 \times 1145 = 5725$. 5 is brought up to the answer.

Another example: $\sqrt{3.000000}$

$$\begin{array}{r} 1. \\ \sqrt{3.000000} \\ \underline{1} \\ 2\ ?\ |\ 200 \end{array}$$

2 goes into 20 10 times but that is too large. 2 goes into 20 7 times; $27 \times 7 = 189$.

$$\begin{array}{r} 1.\ 7 \\ \sqrt{3.000000} \\ \underline{1} \\ 27\ |\ 200 \\ \underline{189} \\ 34?\ |\ 1100 \end{array}$$

Bring down the next pair to the remainder. How many times will 34 go into the first 3 numbers (110)? Approximately 3 times. Change the ? to 3 and multiply by 3: $343 \times 3 = 1029$.

$$
\begin{array}{r}
1.7\ 3 \\
\hline
\sqrt{3.000000} \\
\underline{1} \\
27\ |\ 200 \\
\underline{189} \\
343\ |\ 1100 \\
\underline{1029} \\
346?\ |\ 7100
\end{array}
$$

How many times will 346 go into the first 3 numbers (710)? Approximately 2 times. Change the ? to 2 and multiply by 2: $2 \times 3462 = 6924$.

$$
\begin{array}{r}
1.\ 7\ 3\ 2 \\
\hline
\sqrt{3.000000} \\
\underline{1} \\
27\ |\ 200 \\
\underline{189} \\
343\ |\ 1100 \\
\underline{1029} \\
3462\ |\ 7100 \\
\underline{6924} \\
176
\end{array}
$$

This can be carried out to as many decimal places as needed by continuing the problem. Add a pair of zeros for each decimal needed.

Some square roots are easier to factor first like, $\sqrt{12} = \sqrt{4 \times 3}$. 4 can be easily be factored by its square of $2 = 2\sqrt{3}$ This simply means $2 \times \sqrt{3}$ or $2 \times 1.732 = 3.464$.

Exercise 11-6

1. Find the square root of 135.
2. Find the square root of 180.4.
3. Find the square root of 75 by factoring.
4. Find the square root of A^2.
5. Find the square root of 50.

Solutions to Exercise 11-6

1.
$$\sqrt{135.000000} = 11.618$$

$$20_1\overline{)35}$$
$$\underline{1}$$

$$220_6\overline{)1400}$$
$$\underline{21}$$

$$2320_1\overline{)4400}$$
$$\underline{1356}$$

$$23220_8\overline{)207900}$$
$$\underline{2321}$$

$$\underline{185824}$$
$$22076$$

2.
$$\sqrt{180.4000} = 13.43$$

$$20_3\overline{)80}$$
$$\underline{1}$$

$$260_4\overline{)1140}$$
$$\underline{69}$$

$$2680_3\overline{)8400}$$
$$\underline{1056}$$

$$\underline{8049}$$
$$351$$

3. $\sqrt{75} = \sqrt{25 \times 3} = 5\sqrt{3} = 5 \times 1.732 = 8.66$

4. $A^2 = A \times A$

5.
$$\sqrt{50.000000} = 7.071$$

$$1400_7\overline{)10000}$$
$$\underline{49}$$

$$14140_1\overline{)15100}$$
$$\underline{9849}$$

$$\underline{14141}$$
$$959$$

11-7   METRIC MEASUREMENTS

Temperature Conversions

The metric system is a system of measurement that is used extensively throughout the world. It is very different than our system. We measure temperature in Fahrenheit. Freezing is 32 degrees and boiling is 212 degrees. The metric measurements are in Celsius. Freezing is 0 degrees and boiling is 100 degrees. In Fahrenheit, there are 180 degrees between freezing and boiling. In Celsius, there are 100 degrees between the boiling point and freezing point. That means that for every degree of Celsius there is 1.8 degrees of Fahrenheit. If both Fahrenheit and Celsius started at 0 degrees, then it would just be a matter of multiplying the Celsius temperature by1.8 to convert to the Fahrenheit temperatures, but there is a 32 degree difference in the starting point that must be considered.

To convert 50 degrees Celsius to Fahrenheit, multiply $50 \times 1.8 = 90 + 32 = 122$ degrees. To convert 100 degrees Fahrenheit to Celsius, subtract 32, then multiply by 0.5555 (reciprocal of 1.8): $100 - 32 = 68 \times 0.5555 = 37.8$ degrees.

$$F = C \times 1.8 + 32$$

$$C = F - 32 \times 0.5555 \text{ (reciprocal of 1.8)}$$

An easier way to convert from Celsius to Fahrenheit or Fahrenheit to Celsius is to convert 1.8 to a fraction:

$$\frac{1}{1.8} = \frac{\cancel{10}_5}{\cancel{18}_9} = \frac{5}{9}$$

$$C = \frac{5}{9} \times (F - 32) \quad 100 \text{ degrees } F = (100 - 32) \times \frac{5}{9}$$

$$\text{or } 68 \times \frac{5}{9} = \frac{340}{9} = 37.8 \text{ degrees}$$

$$F = \left(\frac{9}{5} \times C \right) + 32 \quad 50 \text{ degree } C = \left(\frac{9}{5} \times 50 \right) + 32$$

$$\text{or } \frac{4\cancel{50}}{\cancel{5}} \quad 90 + 32 = 122 \text{ degrees}$$

Linear Measurement

Linear measurement in the United States is done in inches, feet, yards, and miles. There are 12 inches in a foot, 36 inches in a yard, and 5280 feet in a mile. The inch is usually divided by 16ths, but can be divided by 10ths.

The metric system is in tens with the meter as the base:

A dekameter is 10 meters

A hectometer is 100 meters

A kilometer is 1000 meters

A Megameter is 1,000,000 meters

On the other side of the decimal:

A decimeter is 0.1 meter

A Centimeter is 0.01 meter

A millimeter is 0.001 meter

A micrometer is 0.000001 meter

A meter is equal to 39.39 inches. To find a decimeter, divide by 10 = 3.939 inches. To find a centimeter, divide a decimeter by 10 = .3939 inches.

To convert from metrics to standard, divide by the unit needed. 70 centimeters = ? inches. 0.3939 inches × 70 = 27.57 inches. 70 centimeters is equal to 7 decimeters or 0.7 meters. 0.7 × 39.39 = 27.57 inches. 0.57 inches is between $\frac{9}{16}$ (0.5625) and $\frac{5}{8}$ (0.6250). 36 inches or 1 yard = ? decimeters. $\frac{36}{3.939}$ or $\frac{36000}{3939}$ = 9.139 decimeters. Divide 9.139 decimeters by 10 = 0.9139 meters.

Weight Measurements

We measure in ounces, pounds and tons. There are 16 ounces in a pound, and 2200 pounds in a ton. In the metric system, the gram is the base measurement. There are 1000 grams in a kilogram. To convert to pounds, there are 454 grams to a pound. 1 kilogram or 1000 grams = ? pounds. $\frac{1000}{454}$ = 2.2 pounds. 1000 kilograms = 1 metric ton. To convert metric tons to pounds, multiply 2.2 × 1000 = 2200 pounds.

A dekagram is 10 grams

A hectogram is 100 grams

A kilogram is 1000 grams

A megagram is1,000,000 grams

On the other side of the decimal:

A decigram is 0.1 gram

A centigram is 0.01 gram

A milligram is 0.001 gram

Liquid Measurement

In standard liquid measurement, fluid ounces, cups, pints, quarts, and gallons are used. There are 32 fluid ounces in a quart. There are 8 fluid ounces to a cup, 2 cups to a pint, 2 pints to a quart, and 4 quarts to a gallon. The metric system of liquid measurement is based on the liter. 1 liter = 1.0567 quarts.

1 dekaliter = 10 liters

1 hectoliter = 100 liters

1 kiloliter = 1000 liters

On the other side of the decimal:

1 deciliter = 0.1 liters

1 centiliter = 0.01 liters

1 milliliters = 0.001 liters

To convert 1 gallon to liters $= \dfrac{4}{1.0567}$ or $\dfrac{40000}{10567} = 3.79$ liters. To convert 6 liters to quarts $= 6 \times 1.0567 = 6.34$ quarts.

Exercise 11-7

1. What is 65 degrees Fahrenheit in Celsius?
2. How many kilograms are in 150 pounds?

3. A transformer holds 4 liters of transformer oil. How many gallons would this be?

4. How many decimeters are in 78 inches?

5. How much is 250 kilograms in pounds?

6. 36 degrees Celsius is what in Fahrenheit?

7. How many liters are in 50 gallons?

8. 50 yards is how many meters?

9. How many centimeters equals 10 inches?

10. A 10 degree rise in Celsius is how much of a rise in Fahrenheit?

Solutions to Exercise 11-7

1. $65 - 32 = 33 \times \dfrac{5}{9} = 18.33$ degrees

2. $\dfrac{150}{2.2}$ or $\dfrac{1500}{22} = 68.18$ kilograms

3. $4 \times 1000 \times 1.0567 = 4226.8$ quarts; $\dfrac{4226.8}{4} = 1056.7$ gallons

4. $\dfrac{78}{3.939}$ or $\dfrac{78,000}{3939} = 19.8$ decimeters

5. $250 \times 2.2 = 550$ pounds

6. $36 \times \dfrac{9}{5} + 32 = \dfrac{\cancel{324}^{64.8}}{\cancel{5}_1} + 32 = 96.8$ degrees

7. $50 \times 4 = \dfrac{200}{1.0567}$ or $\dfrac{2,000,000}{10,567} = 189.27$ liters

8. $50 \times 36 = 1800. \dfrac{1800}{39.39}$ or $\dfrac{1,800,000}{3939} = 45.7$ meters

9. $\dfrac{10}{.3939}$ or $\dfrac{10,000}{3939} = 25.39$ centimeters

10. $10 \times \dfrac{9}{5} = \dfrac{\cancel{90}^{18}}{\cancel{5}_1} = 18$ degrees Fahrenheit

CHAPTER 11 TEST

1. Convert $\dfrac{15}{16}$ to a decimal.

2. What is 6% of 40?

3. Divide $\dfrac{1}{3}$ by $\dfrac{3}{4}$.

4. Solve for X when $\dfrac{3X}{2} = 6$.

5. If the tangent of an angle is 0.4040, what is the angle?

6. What is the square root of 78.47? Do manually.

7. Convert 60 degrees Fahrenheit to Celsius.

8. Multiply 0.7 times 0.03.

9. What is 125% of 16?

10. What is the reciprocal of 4?

11. Solve for X when $X - 4 = 3$.

12. What is the sine of 15 degrees?

13. What is the square root of 0.25?

14. Convert 16 pounds to kilograms.

15. Subtract 16% from 66.

16. Reduce the fraction $\dfrac{64}{256}$.

17. $C^2 = 289$. Solve for C.

18. A board is 45 centimeters long. What is the measurement in inches?

19. What is $\dfrac{1}{2}$ of $\dfrac{3}{4}$?

20. 75% of $X = 12$. What is X?

21. Convert 36 degrees Celsius to Fahrenheit.

22. If the hypotenuse is 7.1 inches and the opposite side is 3 inches, what is the angle?

23. $2X - 4 = 6$. Solve for X

24. How many liters are in 10 gallons?

25. Add $\dfrac{1}{3} + \dfrac{1}{4} + \dfrac{3}{8}$.

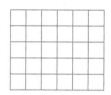

ANSWERS TO
END-OF-CHAPTER TESTS

CHAPTER 1

1. $2 + 2 = 4 \, \dfrac{120}{4} = 30$ amperes

2. $\dfrac{4}{2} = 2 \, \dfrac{120}{2} = 60$ amperes

3. $\dfrac{1800}{10} = 180$ volts

4. $\dfrac{1}{6} + \dfrac{1}{4} + \dfrac{1}{3} = \dfrac{2}{12} + \dfrac{3}{12} + \dfrac{4}{12} = \dfrac{9}{12} = \dfrac{4}{3} = 3\overline{)4} = 1.33$ ohms

5. $E = 100, P = 400.\ \dfrac{E^2}{P} = R.\ \dfrac{10{,}000}{40} = 25$ ohms

6. $\dfrac{6}{3} = 2$ ohms

7. $\dfrac{R_2 \times R_3}{R_2 + R_3} = R_A = \dfrac{8 \times 4}{12} = 2.66.$ Combine $R_5 + R_6 = R_B = 1 + 1 = 2.$
   Combine R_B with $R_7, \dfrac{R}{N} = R_C = \dfrac{2}{2}$ or 1 ohm, $R_1 + R_A + R_4 + R_C + R_8 =$
   $R_T = 2 + 2.66 + 5 + 1 + 10 = 20.66$ ohms.

8. $\dfrac{100}{20.66} = 4.8$ amperes

9. $2.66 \times 4.8 = 12.76$ volts

10. $R_5 + R_6 = R_8 = 1 + 1 = 2.$ Combine R_8 with $R_7, \dfrac{R}{2} = \dfrac{2}{2} = 1$ ohm

11. 4.8 amperes

12. $\dfrac{1}{R_T} = \dfrac{1}{R_1} + \dfrac{1}{R_2} + \dfrac{1}{R_3}$

13. $I_T = I_1 = I_2 = I_3$ 10 amperes

14. $\dfrac{50^2}{5} = \dfrac{2500}{5} = 500$ watts

15. $E_1 = 5 \times 3 = 15$, $E_2 = 5 \times 6 = 30$, $E_3 = 5 \times 8 = 40$, $E_T = 15 + 30 + 40 =$ 85 volts

16. $\sqrt{500 \times 5} = \sqrt{2500} = 50$ volts

17. $\dfrac{50}{5} = 10$ amperes

18. True

19. $\dfrac{50}{5} = 10$ amperes

20. 150 volts

21. $\dfrac{E}{R} = I$. $I_1 = \dfrac{240}{10} = 24$ amperes. $I_2 = \dfrac{240}{20} = 12$ amperes

22. $I_1 + I_2 = I_T$. $24 + 12 = 36$ amperes

23. $\dfrac{40}{4} = 10$ ohms

24. $\dfrac{R}{N} = \dfrac{6}{2} = 3$ ohms. $E = IR = 3 \times 8 = 24$ volts

25. True

CHAPTER 2

1. Yes

2. $240 \times \sqrt{3} \times 10 = 4156$ watts

3. Orange

4. 120 amperes $+ \left(\dfrac{50,000}{795} = 63 \text{ amperes} \right) + 100 \times 360 = \left(\dfrac{36,000}{795} = 45 \right)$
   $= 120 + 63 + 45 = 228$ amperes

5. 1200 kVA $\times$ 0.75 = 900 kW $\dfrac{500}{.90} = 555$ kVA $\sqrt{1200^2 - 900^2} = 793$ kvars; $\sqrt{555^2 - 500^2} = 241$ kvars; 500 kW + 900 kW $= 1400$ kW, 793 kvars + 241 kvars $= 1034$ kvars; $1400^2 + 1034^2 = \sqrt{3029156} = 1740$ kVA; $1400 \div 1740 = 80\%$ power factor.

6. $CM = \dfrac{2KIL \times 866}{VD} = \dfrac{2 \times 12 \times 120 \times 1000}{14.4} \times 866 = 173,200 \; CM =$
   4/0 copper

7. $\sqrt{8^2 + 10^2 + 3^2 - 8 \times 10 - 10 \times 3 - 8 \times 3} = \sqrt{64 + 100 + 9 - 80 - 30 - 24}$
   $= \sqrt{39} = 6.2$ amperes

8. Resistance = 0.510 per thousand feet. $I \times R = E$; $7000 \times 2 = \dfrac{1400}{1000} =$
   1.4 ($700 \times 2 = 1400$). So $1.4 \times 0.510 \times 50 = 35.7$ volts. No

9. $1200 \times 0.85 = 1020$ kW

10. $\sqrt{1200^2 \quad - \quad 1020^2} \quad = \quad 632 \quad$ kvars

11. $\dfrac{1020}{1200} = 0.85 =$ cosine of angle = 31.8 degrees

12. $1500 \times 0.70 = 1050$ kW

    Cosine of 0.70 = 45.57 – tangent = 1.0200
    Cosine of 0.85 = 32 tangent   = –0.6197
                                    0.4003 × 1050 = 420 kvar

13. Cosine of 0.85 = 32 degrees

14. $\dfrac{1050}{0.85} = 1235$ kVA

15. $\dfrac{P}{E} = I = \dfrac{1600}{120} = 13.3$ amperes

16. Phase-to-neutral voltage $= \dfrac{1}{2} \times 220 = 110$; $110 \times \sqrt{3} = 110 \times 1.732 =$
    190.5 volts

17. Since length, amperage, voltage drop, and circular mils are known,
    solve for K.

    $\dfrac{2KIL}{Cm} = VD = \dfrac{2 \times K \times 20 \times 200}{10{,}380 \times K} = \dfrac{13.87}{K}; \dfrac{8000}{10{,}380 \times 13.87} = \dfrac{13.87}{K \times 13.87};$

    $\dfrac{8000}{143{,}970} = \dfrac{1}{K} = \dfrac{143{,}970}{8000} = 18 = 13.87$ volts

    $K = 18$ aluminum

18. B phase

19. $15 - 5 = 10$ amperes

20. $13 \times 1000 = 13{,}000$ volts

21. $\dfrac{P}{E} = I; \dfrac{16{,}000}{240\sqrt{3}} = \dfrac{16{,}000}{415} = 38.55$ amperes

22. $E \times I = P$; $240 \times 32 = 7680$ watts

23. $\dfrac{kW}{kVA} = PF; \dfrac{900}{1059} = 85\%$

24. 32 degrees = cosine of angle = 85%

25. Decreased

CHAPTER 3

1. To step up or step down voltage.
2. $265 \times \sqrt{1.732} = 459$ volts
3. Delta
4. Wye
5. $30 \times 0.866 = 25.9$ kVA
6. $\dfrac{75,000}{360} = 208.3$ amperes
7. $\dfrac{75,000}{240 \times \sqrt{3}} = 181$ amperes
8.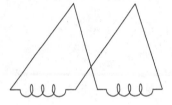
9. $\dfrac{10,000}{240} = 42$ amperes
10. $\dfrac{10,000}{120} = 83$ amperes
11. Adjusts the voltage up or down
12. Subtractive
13. Parallel
14. Series
15. Core loss and copper loss in the transformer windings
16.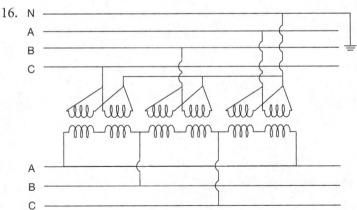

17. $TR = VR$; $30 \times 240 = 7200$ volts
18. 6. Two coils per transformer.
19. 6
20. 2

CHAPTER 4

1. 1 inch. See Chapter 9, Annex C, Table C1 of NEC.
2. #12 = 6 × 0.0133 = 0.0798
   #10 = 6 × 0.0211 = 0.1266 ‾‾‾‾‾‾
    0.2064 = ¾ inch
3. $4\frac{11}{16} \times 2\frac{1}{8}$
4. 26.5 cubic inches
5. 15 × 15
6. 44 × 44
7. The semiconductor fills in between the outer strands and makes the conductor smooth and round. This cuts down the chance of an air pocket forming.
8. 1/0
9. #8
10. 298 amperes
11. 43.3 amperes
12. 24 × 1.75 = 42. Next size breaker up is 50 amperes
13. 40 × 0.89 × 100% = 36
    30 × 0.95 × 85% = 24
    50 × 0.84 × 100% = 42 ‾‾‾‾
     102 amperes = #2 THW copper
14. 200 amperes
15. 60 ampere breaker, 125% of current
16. 3½ inch
17. 3″
18. A
19. #6 copper

20. 28 kW

21. 29,750 watts

22. $2 \times 40 \times 0.71 = 57$
$2 \times 50 \times 0.63 = 63$
$2 \times 30 \times 0.55 = 33$
Largest welder is 29 amperes $\times$ 100% = 29
All other welders = 125 $\times$ 60%   = 75
                                       104 ampers, #2 THW copper

23. 350 amperes

24. 19

25. 5 cubic inches

CHAPTER 5

1. Locked-rotor current is the inrush current required to start the motor.

2. Full-load current is the running current achieved when the motor is up and running.

3. 162 amperes

4. 35 amperes

5. #6 THWN conductors

6. 40 hp = 52 + 2 − 30 hp (40 × 2 = 80) = 52 × 1.25 + 80 = 145 = 1/0 THW copper

7. 52 × 300% = 156 = 175 ampere non-time-delay

8. 52 × 175% + 40 + 40 = 171 = 150 ampere dual-element fuses

9. Short circuit and ground-fault protection

10. 52 × 150% = 78 amperes

11. 14 × 1.25 = 17.5 amperes

12. 1100%

13. 160 amperes

14. 90 amperes

15. Series. Parallel

16. Holding coil

17. True

18. It must be capable of being locked in the open position.

19. 60 ampere disconnect; $52 \times 1.15 = 59.8$

20. Do Not Operate Under Load

21. 60 horsepower; $25 + 52 = 77$ amperes $= 60$ hp

22. $77 \times 1.15 = 84$ amperes $= 100$ ampere disconnect

23. $54 \times 1.25 = 67.5$ amperes

24. 140%

25. Branch circuit short circuit and ground-fault device

CHAPTER 6

1. Four circuits

2. 3850 watts

3. Two

4. 9200 watts

5. 6440 watts

6. 75%

7. 1500 watts

8. None. It is included in the general lighting and receptacle load.

9. Heating and air conditioning

10. 10 kW

11. Existing load and new load are added together.

12. First 8 kW at 100%, remainder at 40%

13.

Lighting: 3 watts per sq ft	$1800 \times 3 =$	5400
Appliance circuits	$1500 \times 2 =$	3000
Laundry circuit	$1500 \times 1 =$	1500
Oven	12,000	12,000
Clothes Dryer	5000	5000
Dishwasher	600	600
Garbage Disposal	500	500
Water heater	3000	3000
Attic fan	250	250
Total		31250
	$-\underline{10,000} \times 100\%$	$=$	10,000
	$21250 \times 40\%$	$=$	8500
Subtotal		$=$	18,500

Heat	10,000		10,000
			28,500

$$\frac{28,500}{240} = 119 \text{ amperes}$$

14.

Lighting: 3 watts per sq ft	$1800 \times 3 =$	5400	
Appliance circuits	$1500 \times 2 =$	3000	
Laundry circuit	$1500 \times 1 =$	1500	
		9900	
		$- 3000 \times 100\%$	3000
		$6900 \times 35\%$	2415
Oven	12,000	$8000 \times 70\%$	5600
Clothes dryer	5000	$5000 \times 70\%$	3500
Dishwasher	600	$600 \times 75\%$	450
Garbage disposal	500	$500 \times 75\%$	375
Attic fan	250	$250 \times 75\%$	188
Total			15,528

$$\frac{15,528}{240} = 65 \text{ amperes}$$

15. 100 amperes

16. Two #2, one #4 copper conductors

17.

Lighting: 3 watts per sq ft	$1500 \times 3 =$	4500	
Appliance circuits	$1500 \times 2 =$	3000	
Laundry circuit	$1500 \times 1 =$	1500	
		9000	
		$- 3000 \times 100\% =$	3000
		$6000 \times \ 35\% =$	2100
			5100
Oven	12,000	8000	8000
Clothes dryer	5000		5000
Dishwasher	600	$\times 75\%$	$= \quad 450$
Garbage disposal	700	$\times 75\%$	$= \quad 525$
Water heater	3000	$\times 75\%$	$= \quad 2250$
Attic fan	250	$\times 75\%$	$= \quad 188$
Heat pump	10,000		10,000
Total			31,513

$$\frac{31,513}{240} = 131 \text{ amperes}$$

18.

Lighting: 3 watts per sq ft		$1500 \times 3 =$	4500
Appliance circuits		$1500 \times 2 =$	3000
Laundry circuit		$1500 \times 1 =$	1500
			9000

$$-\underline{3000} \times 100\% = 3000$$
$$\overline{6000} \times 35\% = \underline{2100}$$
$$5100$$

Oven	12,000	$8000 \times 70\%$		5600
Clothes dryer	5000	$5000 \times 70\%$		3500
Dishwasher	600	$\times 75\%$	$=$	450
Garbage disposal	700	$\times 75\%$	$=$	525
Attic fan	250	$\times 75\%$	$=$	188
Total				15,363

$$\frac{15,363}{240} = 64 \text{ amperes}$$

19. Two 2/0, one #4 aluminum

20. True

CHAPTER 7

1. 100 amperes

2. No if laundry is not allowed or if laundry services are provided on the premises and are available to all occupants.

3. Table 220.44, column B: $7000 \times 0.36 = 2520$ watts

4. Table 220.61: $2520 \times 0.70 = 1764$ watts

5. Table 220.55, column B: $7000 \times 0.80 = 5600$ watts

6. Table 220.61: $5600 \times 0.70 = 3920$ watts

7. Table 220.54: % = 47 − (number of dryers − 11)
   % = 47 − (16 − 11) = 47 − 5 = 42%
   42% × 5000 × 16 = 33,600 watts

8. Table 220.61: $33,600 \times 70\% = 23,520$ watts

9.

			Phase	Neutral
Lighting: 3 watts per sq ft	$600 \times 3 =$	1800		

					Phase	Neutral
Appliance circuits		$1500 \times 2 =$	3000			
			4800			
			$-\underline{3000} \times 100\%$	$=$	3000	3000
			$1800 \times 35\%$	$=$	$\underline{630}$	630
					3630	
Oven	8000	$\times 80\%$	6400		6400	
70% for Neutral			$6400 \times 70\%$			4480
Dishwasher	600			$=$	600	600
Garbage disposal	750			$=$	750	750
Water heater	2500			$=$	2500	
Total					13,880	9460

Phase	Neutral
$\dfrac{13,880}{240} = 58$ amperes	$\dfrac{9460}{240} = 39$ amperes

58 amperes for the phases; 39 amperes for the neutral.

10.

					Phase	Neutral
Lighting: 3 watts per sq ft		$600 \times 3 \times 8$	14,400			
Appliance circuits		$1500 \times 2 \times 8$	24,000			
			38,400			
			$-\underline{3000} \times 100\%$	$=$	3000	3000
			$35,400 \times 35\%$	$=$	$\underline{12,390}$	12,390
					15,390	
Oven	8000	$\times 8$	$64,000 \times 36\%$		23,040	
70% for Neutral			$23,040 \times 70\%$			16,128
Dishwasher	600	$\times 8 \times 75\%$	3600	$=$	3600	3600
Garbage disposal	750	$\times 8 \times 75\%$	4500	$=$	4500	4500
Water heater	2500	$\times 8 \times 75\%$	15,000	$=$	15,000	
Total					61,530	39,618

Phase	Neutral
$\dfrac{61,530}{240} = 256$ amperes	$\dfrac{39,618}{240} = 165$ amperes

11.

Lighting: 3 watts per sq ft		$700 \times 3 \times$	32	67,200
Appliance circuits		$1500 \times 2 \times$	32	96,000

Oven	8000	×	32	256,000
Dishwasher	600	×	32	19,200
Garbage disposal	500	×	32	16,000
Heater	2250	×	32	72,000
Water heater	3000	×	32	96,000
Total				622,400

$$622,400 \times 31\% \quad \frac{192,944}{240} = 804 \text{ amperes}$$

12.

Lighting: 3 watts per sq ft	$700 \times 3 \times$	32	67,200
Appliance circuits	$1500 \times 2 \times$	32	96,000
			163,200

$$\frac{-3000}{160,200} \times 100\% = \quad 3000$$

$$\frac{-117,000}{43,200} \times 35\% \quad 40,950$$
$$43,200 \times 25\% \quad 10,800$$

Oven	8000	× 0.22	32	56,320		
70% for Neutral				56,320 × 70%		39,424
Dishwasher	600	× 0.75	32	14,400	=	14,400
Garbage disposal	500	× 0.75	32	12,000	=	12,000
Heater	2250	×	32	72,000		72,000
Water heater	2500	× 0.75	32	60,000	=	
Total						192,574

Neutral

$$\frac{192,574}{240} = \quad 802 \text{ amperes}$$

$$\begin{array}{r} 802 \\ -200 \\ \hline 602 \times \end{array} \quad 70\% = \quad \begin{array}{r} 200 \\ 421 \\ \hline 621 \text{ amperes} \end{array}$$

13. House loads are loads that are common to all occupants, such as laundry rooms, pools, office, outside lighting, and other amenities.

14. No. House loads must be calculated separately and added to the load.

15.

$$450 \times 20 \times 3 = 27,000$$

$$\text{2 appliance circuits: } 2 \times 1500 \times 20 = \frac{60,000}{87,000}$$

$$\frac{-3000}{84,000} \times 100\% = \quad 3000$$

$$84,000 \times 35\% \quad = \frac{29,400}{32,400 \text{ watts}}$$

16. $87,000 \times 38\% = 33,060$ watts

17. Yes, if a cooking unit of 8 kW per unit is used. This can be compared to the standard calculation without a cooking unit, and the smaller of the two calculations may be used.

18. $3 \times 12 = 36$
    $4 \times 14 = 56$
    $3 \times 15 = \underline{45}$
    $137/10 = 13.7$ per unit. Average range is 14 kW or 8800 watts.

19. $8800 \times 70\% = 6160$ watts

20. $30 \times 150 \times 125\% = \dfrac{5625}{240} = 23$ amperes

21. $5000 \times 4 = 20,000$ watts

22. $20,000 \times 0.45 = 9000$ watts

23. $900 \times 3 = \dfrac{2700}{120} = \dfrac{22.5}{15} = 2$ circuits

24. Table 220.55: 39 kW

25. $39,000 \times 0.70 = 27,300$ watts

CHAPTER 8

1. $25,000 \times 1 = 25,000$
   $- \underline{10,000} \times 100\% = 10,000$
   $15,000 \times 50\% = \underline{7500}$
   $17,500$ watts

2. 80% (Table 220.56)

3. $25,000 \times 3.5 \times 125\% = 109,375$ watts

4. 125,000
   $\underline{50,000} \times 40\% = 20,000$
   $75,000 \times 20\% = \underline{15,000}$
   $35,000$ watts

5. A load in which the maximum current is expected to be on for 3 hours or more.

6. 1200 watts (Article 600.5A and Article 220.3B6)

7. 200 watts a linear foot: $200 \times 60 \times 125\% = 15,000$ watts

8. $3\frac{1}{2} \times 1500 \times 125\% = 6563$ watts
   $\frac{1}{4} \times 3000 \times 125\% = \underline{938}$ watts
   $\phantom{\frac{1}{4} \times 3000 \times 125\% = }7501$ watts

9. $25,000 \times 0.25 = 6250$ watts

10. 280 amperes, no deration allowed

11. 25%

12. 180 watts

13.   325,000
    − 200,000   = 160,000
     125,000 × 10% = 12,500
                      172,500 watts

14.   325
    − 200 × 100% = 200,000
     125 × 25% = 62,500
                262,500 watts

15. $600 × 125\% = \dfrac{750}{2} = 375$ amperes = six 500 MCM THW copper conductors

    $300 × 125\% = \dfrac{375}{2} = 188$ amperes = two 3/0 THW copper conductors

16. $0.7901 × 3 = 2.3703$
    $0.3117 × 1 = 0.3117$
    $0.1901 × 1 = \underline{0.1901}$
                2.8721 = two 3 inch rigid conduits

17. 2/0 copper (Table 250. 66)

18. $22 × 1.40 × 6 × 0.79 = 146$ amperes

19. $150 × 360 = \dfrac{54000}{795} = 68$ amperes

20. $22 × 2.50 + (5 × 22) = 165$ amperes. Next-lower breaker = 150 amperes.

21. $200 × 125\% × 125 = \dfrac{31{,}250}{277} = \dfrac{113}{20} = 6$ circuits

22. $3.5 × 15{,}000 × 125\% = 65{,}625$ watts. No. The service, panels, and feeders must be sized by the larger of the two values, but the actual branch circuits for the design value may be used. 125 200 watt fixtures would require six circuits ($\dfrac{125 × 200 × 1.25}{277 × 20} = 6$)

23. $200 × 125\% = 250; 277 × 20 = \dfrac{5540}{250} = 22$ fixtures

24. $\dfrac{15{,}000}{240} = 63$ amperes

25. $\dfrac{15{,}000}{360} = 42$ amperes

CHAPTER 9

1. $\dfrac{2,500,000}{50,000} = 50$ volt amperes per square foot.

$$
\begin{array}{r}
50 \\
\underline{-\ 3} \times 100\% \times 50,000 = 150,000 \\
43 \\
\underline{-\ 17} \times 75\% \times 50,000 \ = 637,500 \\
30 \times 25\% \times 50,000 \ = 375,000 \\
\overline{1,162,000 \text{ watts}}
\end{array}
$$

2. $\dfrac{1,162,000}{795} = 1462$ amperes

3.
$$
\begin{array}{l}
150 \\
\underline{-\ 60} \times 100\% = 60 \\
90 \\
\underline{-\ 60} \times 50\% \ \ = 30 \\
30 \times 25\% \ \ = \ 7.5 \\
\overline{97.5}
\end{array}
$$

First two buildings = 98 amperes × 100% = 98
Building 3 = 80 amperes × 75% = 60
Building 4 = 50 amperes × 65% = 33
Building 5 = 30 amperes × 50% = 15
Dwelling = 108
313 amperes

Dwelling: $\dfrac{26,000}{240} = 108$ amperes

4.

			Phase	Phase
Lighting: 50 × 8 × 3 watts per sq ft		400 × 3	1200	
Appliance circuits		1500 × 2	3000	
Laundry circuit		1500 × 1	1500	
			5700	
		−3000 × 100% =	3000	3000
		2700 × 35%	945	945
Oven	6000	× 0.80	0	4800	4800
Blower fan	250				250
Water heater	1500	×	0	= 1500	
Total				10,245	8995

The largest leg is 10,245/240 = 43 amperes 50 ampere power cord.

5. To provide the number of poles, voltage, amperage, and frequency of the intended service.

6. Three

7. Minimum 21 feet; maximum 36.5 feet

8. 8000 volt amperes

9. 100 amperes

10. $16{,}000 \times 6 \times 0.29 = \dfrac{27{,}840}{240} = 116$ amperes

11. 125 volts

12. 15, 20, 30, and 50 amperes

13. 50 amperes = 5; 30 amperes = 70

14. None

15. 9600 volt amperes

16. $\begin{aligned} 9600 \times 5 &= 48{,}000 \\ 3600 \times 35 &= 126{,}000 \\ 2400 \times 5 &= 12{,}000 \\ 600 \times 5 &= 3000 \\ \hline 189{,}000 \times 0.41 &= \dfrac{77{,}490}{240} = 323 \text{ amperes} \end{aligned}$

17. $\begin{aligned} 150{,}000 \times 2 &= 300{,}000 \\ -20{,}000 \times 50\% &= 10{,}000 \\ \hline 280{,}000 \\ -80{,}000 \times 40\% &= 32{,}000 \\ \hline 200{,}000 \times 30\% &= 60{,}000 \\ \hline 102{,}000 \text{ watts} \end{aligned}$

18. 87%

19. $\begin{aligned} 10{,}000 \\ 4375 \\ \hline 14{,}375 \end{aligned}$

    $\dfrac{14{,}375}{277} = \dfrac{52}{20} = 3$ circuits

20. $\dfrac{15{,}000}{480} = \dfrac{31}{20} = 2$ circuits

21. $150{,}000 \times 2 = \dfrac{300{,}000}{360} = \dfrac{833}{20} = 42$ circuits

22. $\dfrac{56{,}405}{795} = 71$ amperes

23. $\dfrac{35{,}861}{360} = 100$ amperes

24. $124 \times 125\% = 155$ amperes = 2/0 THW copper conductors

25. $124 \times 175\% = 217$ amperes; nearest standard size is 225 ampere.

CHAPTER 10

1. 0.6645 or approximately ⅝ inches
2. length – gain + shrinkage of offset: 55 – 3½ + 1⅝ = 49⅞ inches
3. ½ tangent of angle × spacing + outside diameter of conduit. B = 8 + 1.11 = 9.11; C = 9.11 + 1.11 = 10.22 inches
4. $^5/_{15}$ = 0.3333 = 19.46 degrees
5. 2½ × 3 = 7½ inches
6. 36 + 15 + 15 – 7 = 59 inches
7. 10 × 0.414 = 4.14 inches; 60 – 4.14 = 55.86 inches
8. 8 × 2 + ½ of 2⅛ = 16 + 11¹/₁₆ = 17¹/₁₆ inches
9. 17¹/₁₆ + 4 = 21¹/₁₆ inches
10. 21¹/₁₆ × 1.57 = 33 inches
11. 36 + 48 + 26.78 – 34.125 = 76.66 inches
12. 40 + 52 + 33 – 42⅛ = 82⅞ inches
13. 1 = 45 degrees; 2 = 22 ½ degrees
14. 45 degrees = 3(½ × 0.414 + 0.1989) = 1.22 inches
15. 2.366
16. 2.144
17. 0.2215 per inch
18. Hypotenuse = 12 inches, opposite side = 6 inches
19. Adjacent side = $\sqrt{12^2 - 6^2}$ = 144 – 36 = 108 = 10.39 inches
20. 12 – 10.39 = 1.61 inches
21. $\dfrac{15}{2\pi}$ = R = 2.388 × 12 = 28.66 inches + 6 = 34.66 inches: $2\pi R = C = 2$ × 3.14 × 34.66 = 217.66 inches or $\dfrac{217.66}{12}$ = 18.14 feet
22. $\dfrac{217.66}{360}$ = 0.605 inches
23. 90 × 0.605 = 54.45 inches
24. 2 × 0.61 = 1.22 inches
25. 20 + 18 + 24 + 36 – 10.5 = 87.5 inches

CHAPTER 11

1. $16\overline{)15.0000}$ quotient 0.9325

2. $40 \times 0.06 = 2.4$

3. $\dfrac{1}{3} \div \dfrac{3}{4} = \dfrac{1}{3} \times \dfrac{4}{3} = \dfrac{4}{9}$ or 0.4444

4. $\dfrac{3X \times 2_1}{2_1} = 6 \times 2 = 3X = 12 \dfrac{3_1 X}{3_1} = \dfrac{12_4}{3_1} = X = 4$

5. 22 degrees

6. $\sqrt{78.470000}$ → $8.\ 8\ 5\ 8$

   $160_8\overline{)14\ 47}$ with 64

   $1760_5\overline{)1\ 03\ 00}$ with $13\ 44$

   $1770_8\overline{)1475\ 00}$ with 8825

   $1416\ 64$

   $58\ 36$

7. $60 - 32 \times \dfrac{5}{9} = 28 \times \dfrac{5}{9} = \dfrac{140}{9} = 15.55$

8. 0.021

9. $16 \times 1.25 = 20$

10. ¼ or 0.25

11. $X - 4 + 4 = 3 + 4 = X = 7$

12. 0.2588

13. 0.5

14. $16 \div 2.2 = 7.264$ kilograms

15. $66 \times 0.16 = 10.56$; $66 - 10.56 = 55.44$

16. $\dfrac{64_1}{256_4} = \dfrac{1}{4}$

17. $C = \sqrt{289} = 17$

18. $45 \times 0.3939 = 17.72$ inches

19. $\dfrac{1}{2} \times \dfrac{3}{4} = \dfrac{3}{8}$

20. $0.75 = 12; \dfrac{^{10}\cancel{0.75}X}{_1\cancel{0.75}} = \dfrac{12}{0.75}; X = 16$

21. $36 \times \dfrac{9}{5} = \dfrac{324}{5} = 64.8 + 32 = 96.8$ degrees

22. $\dfrac{\text{hypotenuse}}{\text{opposite}} = \text{cosecant} = \dfrac{7.1}{3} = 2.3666 = 25$ degrees

23. $\dfrac{^2\cancel{2}_1 X}{\cancel{2}_1} - \cancel{4}_1 + \cancel{4}_1 \dfrac{6+4}{2} = 5$

24. 1 liter = 1.0567 quarts; $10 \times 4 = \dfrac{40}{1.0567} = 37.85$ liters.

25. $\dfrac{1}{3} + \dfrac{1}{4} + \dfrac{3}{8} = \dfrac{8}{24} + \dfrac{6}{24} + \dfrac{9}{24} = \dfrac{^{21}\cancel{21}_7}{\cancel{24}_8} = \dfrac{7}{8}$

INDEX

ABOUT THE AUTHOR

Nick Fowler has been a practicing electrician for 35 years. He holds several state and Master Electrician licenses, and has been an electrical journeyman, foreman, and general foreman. He also taught electrical technology at a community college. Mr. Fowler is a member of International Brotherhood of Electrical Workers Local 191, Everett, Washington, and has worked around the United States as part of the I.B.E.W. His home is currently in San Antonio, Texas.